Agronomy of grassland systems

C.J.PEARSON

and

R.L.ISON *

School of Crop Sciences, University of Sydney, Australia
* Formerly at School of Agriculture, Hawkesbury Agricultural College, Richmond, Australia

The right of the
University of Cambridge
to print and sell
all manner of books
was granted by
Henry VIII in 1534.
The University has printed
and published continuously
since 1584.

CAMBRIDGE UNIVERSITY PRESS

Cambridge

New York   New Rochelle   Melbourne   Sydney

Published by the Press Syndicate of the University of Cambridge
The Pitt Building, Trumpington Street, Cambridge CB2 1RP
32 East 57th Street, New York, NY 10022, USA
10 Stamford Road, Oakleigh, Melbourne 3166, Australia

© Cambridge University Press 1987

First published 1987

Printed in Great Britain

British Library cataloguing in publication data
Pearson, C. J.
Agronomy of grassland systems.
1. Meadows   2. Pastures   3. Stock-ranges
I. Title   II. Ison, R. L.
360'.915'3   SB197

Library of Congress cataloguing in publication data
Pearson, C. J.
Agronomy of grassland systems.
Bibliography
Includes index.
1. Pastures.   2. Grasses.   3. Range management.
I. Ison, R. L.   II. Title.
SB199.P37   1987   633.2   86-33443

ISBN 0 521 32448 3 hard covers
ISBN 0 521 31009 1 paperback

CONTENTS

PREFACE

Television and popular books have made the public aware of the beauty and complexity of the world's great natural grasslands: the Eurasian steppes, the African rift valley and veldt, the Australian savanna, the Argentinian pampas and the North American plains. Less publicized, but more important, is the fact that farmers and grassland agronomists have created systems of smaller area but greater productivity than the natural grasslands. In many countries these grassland systems utilize more solar energy and employ more people than crop enterprises: in Australia, grasslands produce at least three times the vegetation of croplands (R. M. Gifford, personal communication, 1985). The efficiency with which grassland systems convert this energy into meat, milk and fibre is low. Nonetheless, livestock are often the only practical means of converting solar energy into products which are useful to man: grassland agronomy is essential to the provision of livestock products and less measurable benefits such as recreation and wilderness areas and to the control of erosion.

The world population, which is now just over 2 billion, will at current rates of increase rise by the year 2000 to 3 billion. This, together with recurring shortages of food in the 1970s and the realization that pastoralism is energetically less efficient than crop production, led some people to conclude that grasslands would be displaced by cropland. This is happening only in relatively small areas which have a reliable climate and favourable soils. More generally, pastoralism and cropping are in dynamic balance so that commodity prices, not biological efficiency, determine the balance between grassland and cropland within particular regions. In the current decade the over-supply of small grains is pushing the balance towards grasslands.

We both receive satisfaction from understanding, albeit incompletely, how grassland systems work, and we derive excitement from the challenge of trying to make them work 'better'. We hope that this book will give satisfaction and excitement to an audience which is far larger than we encounter personally in our learning and university and college teaching. With a geographically widespread audience in mind we have emphasized systems concepts and biological principles; we have avoided prescriptions which usually have only local relevance. We have also attempted to maintain geographical balance in our examples. Nonetheless, Australia and New Zealand are over-represented in this respect because their contribution to grassland research is disproportionately large in relation to their population and South America and eastern Europe are under-represented because the research findings from these areas were not easily accessible to us.

We thank our wives June and Catherine for their support and June Pearson for word processing. We also thank our colleagues, particularly A. C. Andrews, A. C. Archer, L. C. Campbell, R. D. Davis, J. Donnelly, E. F. Henzell, M. J. Hill, Z. Hochman, L. R. Humphreys, R. M. Jones, R. Kellaway, A. Lazenby, P. Martin, L. Mohammed, J. Mott, A. Muir, R. Packham, I. Valentine and the staff of the Hawkesbury Agricultural College Resources Center for their help.

C. J. Pearson
R. L. Ison

Sydney
March 1986

1

An overview

Grassland agronomy is the science of grassland production. A grassland agronomist may be defined as an analyst who identifies and solves problems of grassland production. These problems and the ways in which the grassland agronomist works to resolve them occur within a farming system. There is a hierarchy of levels of biological organization and of problem-solving strategies within a farming system (Fig. 1.1).

Problem definition (and possibly redefinition) requires care and consideration. To this end, methodologies for defining problems for research have been developed (e.g. Bawden *et al.*, 1985; Conway, 1985*a*). Some problems may be defined in a way which is quite specific, e.g. the poor growth response of plants to a non-optimal condition such as increasing aluminium toxicity in the soil solution. Such problems are amenable to analysis using hypotheses which are testable. The analysis should lead to a definite solution or, at least, to a 'best' or an optimal solution to that specific problem. At the other extreme, in terms of levels of complexity, are problems which arise because of the uncertain interrelationships which are a part of farming systems: there is generally no single, static 'best' solution to problems of interrelationships in a dynamic system involving environment, plants, animals, technology (types of ploughs, etc.), economics and the social values and goals of the farmer. The agronomist thus works within a continuum which ranges from the deduction of solutions to specific biological problems to the consideration of how to make the whole system more satisfying (Fig. 1.1).

If students of agronomy recognize that they are working within this continuum, and farmers or pastoralists recognize that they must make decisions within a system which is so complex that there may be no single, universal best solution, then it is likely that each professional will more fully understand and

appreciate the other. From this recognition it also follows that prescriptive advice from a textbook is likely to have very limited applicability. In this book we try to avoid prescriptions: we are concerned with principles.

Most of this book is devoted to the principles which underly the biological operation of the grassland system, particularly those principles which relate to the dynamics of the biological system. However, farmers have a diversity of forage sources from which they may draw to sustain animal production, so that,

Fig. 1.1. Schema of a grassland system within which the agronomist works. The scheme illustrates the transition from specific hard problems related to biological components of the system to management problems of the 'soft system'. (Adapted from Bawden *et al.*, 1985.)

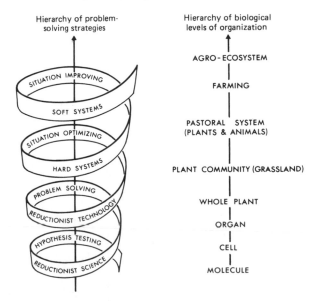

Hierarchy of problem-solving strategies

- SITUATION IMPROVING
- SOFT SYSTEMS
- SITUATION OPTIMIZING
- HARD SYSTEMS
- PROBLEM SOLVING
- REDUCTIONIST TECHNOLOGY
- HYPOTHESIS TESTING
- REDUCTIONIST SCIENCE

Hierarchy of biological levels of organization

- AGRO-ECOSYSTEM
- FARMING
- PASTORAL SYSTEM (PLANTS & ANIMALS)
- PLANT COMMUNITY (GRASSLAND)
- WHOLE PLANT
- ORGAN
- CELL
- MOLECULE

Table 1.1. *Level of control which farmers are able to exercise over environmental and biological variables in pasture systems*

Environmental factor	Level of control	Method of control
Grazing pressure	Good	Stocking rate, stock movement and herd (population) structure
Plant population	Good	Sowing, selective herbicides, stocking rate and stock movement
Defoliation	Good	Stocking rate and stock movement
Nutrition	Fair	Fertilizer application where economic or where species respond
Pests and diseases	Variable, usually poor	Ranges from good short-term control in intensive situations to control relying on plant resistance in extensive systems
Moisture	Poor	Irrigation: selection between existing wetter and drier sites and use of cultivars which have development patterns to match available soil water; farmer can use mechanical aids (contour furrows, drains, etc.)
Soil structure	Poor	Tillage and stocking affect particle size distribution, etc. but knowledge is empirical (e.g. don't plough when soil is 'too wet'): farmer can select between existing soils, and make amendments, e.g. add gypsum
Temperature	Indirect	Selection of sites with different aspects; selection of species with different growth responses to temperature; or alteration of time of sowing

Fig. 1.2. The main biological components in the function and management of grassland systems. A grassland system is dynamic: various pools or state variables (□) are linked by flows of material, e.g. seed, leaf (arrows) and governed by rate variables (▭◁).

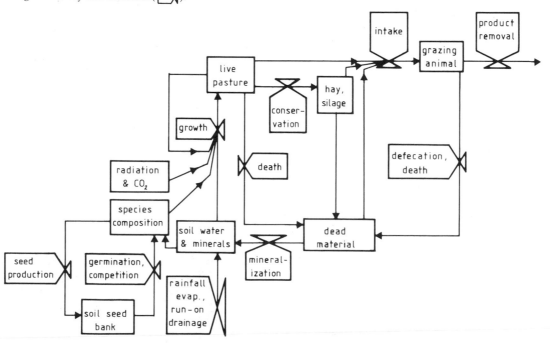

increasingly, grassland agronomists will be concerned with the integration of forage sources, including crop and agro-industrial residues, into animal production systems. Accordingly, some economic principles and likely future roles of grasslands in farming systems are discussed in Chapter 8.

1.1
Biological components of grassland systems

The biological components of grassland or pastoral systems are the environment, plants and animals. These are closely interrelated in a cyclical fashion (Fig. 1.2). We emphasize the quantitative interrelationships within the biological cycle and the feedback nature of the cycle. This contrasts with the viewpoint of the traditional agronomist, who usually thinks of the cycle as a simple catena: environment →growth of grass →animal production →product removal. It is useful to note, in passing, that the amount of control which the farmer is able to exercise over the components in the grassland system ranges from virtually nil to total (Table 1.1).

Notwithstanding our emphasis on a systems view of grassland agronomy, we are bound by the fact that books start at the beginning and end at the last page, to structure this book in a catenary fashion.

The first part of the catena is generation (Fig. 1.2, Chapter 2). This comprises the dynamics of the bank of seed in the soil, seed germination and vegetative generation from stolons and rhizomes, leading to plant emergence. Generation leads to vegetative growth (Chapter 3) and the life cycle of grassland plants ends with seed production (Chapter 4). Nutrition (Chapter 5) links the plant and animal components of the system. Schemata such as that shown in Fig. 1.2 can be made specific by emphasizing particular developmental aspects, environmental variables or loops feeding back material or information. An example of this specificity is provided by a grassland comprised of a mixture of the tropical annual legume Townsville stylo (*Stylosanthes humilis*) and annual grasses in the Australian wet-and-dry tropics. Here, in a climate in which rainfall during the wet season dominates the productivity and composition of the grassland, the agronomist pays particular attention to estimating how the composition and productivity are affected by water through rainfall, infiltration, run-off, soil moisture, soil drying and drought (Fig. 1.3).

The quality and quantity of feed available from living, dead and conserved pools determines animal intake (Chapter 6). The animal in turn affects the productivity and composition of the grassland (Chapter 7).

Finally, the agronomist, economist and farmer integrate the principles of pasture development, growth and utilization into the management of the grassland system. Management is associated with farmers' goals, which vary within socio-cultural systems and between them; where management interventions are made to improve the quantity or quality of herbage (e.g. saving, fertilizing), there is a need for managers to estimate the likely annual cycle of production of grasslands and the requirements of the farm livestock. These must be matched, making allowance if necessary for the conservation or purchase of feed, in a way which ensures some return on the initial investment. Such returns are usually economic (Chapter 8), but in some societies they may be social or cultural rewards. Today we have many tools which we

Fig. 1.3. Environmental and plant factors that dominate grassland pattern and species cycling in a grassland comprised of Townsville stylo and annual grasses in the wet-and-dry tropics. The factors exert quantitative effects on yield and species composition; the plant factors germination, establishment, competition and seed production may be seen as a series of filters through which individual plants attempt to pass. (From Torssell, 1973.)

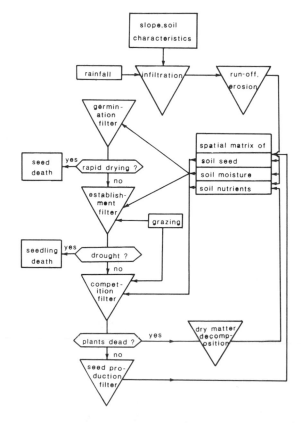

use for estimating production within grassland systems: they range from the valuable but indefinable judgement of experienced farmers to feed year budgeting to probabilistic modelling on mainframe computers.

The output from a grassland system is measured in units of meat, milk, wool, hide or money. This is called production. In this book we define productivity (Fig. 1.4a) as the output per unit of time, e.g. kg beef per year or kg beef per ha per year. Productivity is a 'rate variable', i.e. a measure of the dynamic nature of the system, or how it operates. Rate variables contrast to state variables such as the amount of standing feed, which tell us only what a particular grassland system looks like at a particular point in time.

Defining the units of biological output from a grassland as kg of beef etc. does not, however, indicate, even in biological terms, whether the system is operating successfully or efficiently. Success is usually measured in terms of the short-term (seasonal or annual) output relative to inputs. Here we call this a measure of the efficiency of grassland production (Fig. 1.4b). Of equal importance is the stability of the system, i.e. its ability to return to an 'equilibrium state' after a temporary disturbance (Fig. 1.4c). Finally, some ecologists would differentiate sustainability, i.e. the ability of a system to maintain itself or the degree of difficulty of management required to maintain it.

1.2
Grassland productivity

The net primary productivity (NPP) is the net change in weight of grassland plants between any two points in time, usually over a year. NPP depends on (i) the type of grassland, (ii) the climate and soil and (iii) management.

The diversity of grasslands around the world and the recent intensification of our use of them have been

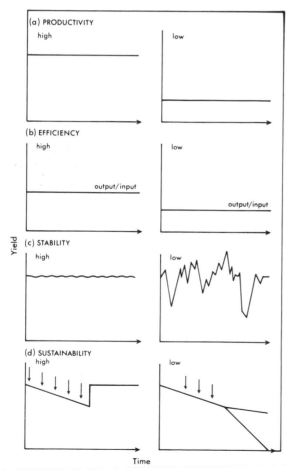

Fig. 1.4. The properties of grassland or pastoral systems defining generalized situations of high (left-hand column) and low (right-hand column) productivity (a), efficiency (b), stability (c) and sustainability (d). In (d) the arrows represent stress imposed on the system. (Adapted from Conway, 1985*b*.)

Table 1.2. *Estimates of NPP and total net production of all vegetation on land areas of 10° latitudinal belts*

Latitudinal belt	NPP (t/(ha year)	Net production (10⁹ t/year)
Northern hemisphere (°N)		
90–80	0.40	0.0
80–70	1.52	0.5
70–60	4.06	5.4
60–50	6.58	9.6
50–40	7.53	12.5
40–30	8.14	12.7
30–20	7.43	11.3
20–10	10.65	12.1
10–0	17.80	18.0
Southern hemisphere (°S)		
10–0	19.68	20.6
20–10	14.07	13.3
30–20	9.00	8.4
40–30	8.93	3.6
50–40	8.27	0.8
60–50	7.83	0.2
70–60	1.28	0.3
80–70	0.12	0.1
90–80	—	—
Total production	—	129.4
Mean NPP	8.66	—

Source: Box (1978).

reviewed by Barnard & Frankel (1966). In the northern hemisphere most of the grasslands are in cool temperate climates; in the southern hemisphere the greatest area is occupied by the savanna grasslands in the wet-and-dry tropics of Africa, South America and Australia (Fig. 1.5). Utilization, however, is greatest in regions which have a wet temperate or mediterranean climate.

Leigh (1973) constructed a world map of NPP by fitting NPP (in g per m^2 per year) to the mean annual temperature T (°C) and the average annual rainfall R (mm):

$$NPP = 3000[1 + \exp(1.315 - 0.119T)] \qquad (1.1)$$

and

$$NPP = 3000[1 - \exp(-0.000\,664R)] \qquad (1.2)$$

Modifications of this model produce values of NPP per unit area of land for regions of the earth's surface (Table 1.2). This and other models show that productivity is highest in the tropics: the NPP is likely to be of the order of 9 t per ha per year in those regions where grasslands are the predominant use of farmland.

1.3
Efficiency

A holistic view of grassland agronomy implies that there are many ways of measuring the efficiency of a system. Moreover, the dynamic nature of grasslands makes it difficult to be sure what one is, or should be, comparing in order to calculate efficiency. A change in one variable will incur a direct cost but it will usually also change other variables which in turn have associated costs so that an annual analysis of marginal return may be superficial and one may be unable to apportion returns and costs to specific innovations. For example, if a farmer increases his use of fertilizers this will incur a direct cost but it will also enable him to increase the stocking rate so that realistic analysis of the economic efficiency of the first innovation must relate the return from extra beef with the costs of extra fertilizer, stock, labour, cartage, etc.

Efficiency is most meaningful when assessed in terms of transformations within the system. There are several such measures of efficiency:

(i) Efficiency of NPP. This expresses the energy fixed in NPP as a percentage of the energy of solar radiation which is incident on the grassland (Section 3.6).

(ii) Efficiency of recovery of fertilizer. This is the amount of extra mineral element which is recovered in herbage expressed as a percentage of the amount of additional fertilizer applied to the soil (Section 5.1).

Fig. 1.5. Distribution of native grasslands and modified, including sown, grasslands. Unhatched areas are not predominantly grasslands. (From Moore, 1966.)

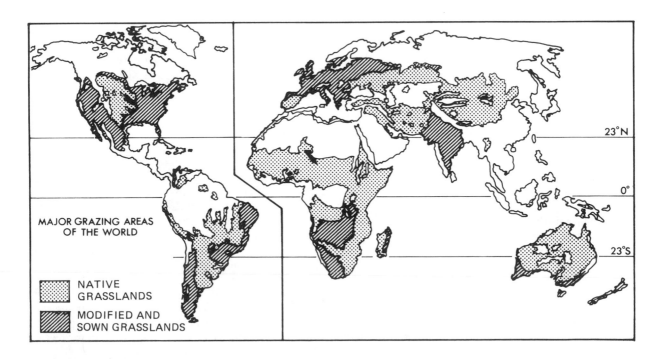

MAJOR GRAZING AREAS OF THE WORLD

NATIVE GRASSLANDS

MODIFIED AND SOWN GRASSLANDS

23°N

0°

23°S

(iii) Efficiency of grazing or harvesting. This may be expressed as (a) the herbage consumed at each grazing as a proportion of the herbage mass present or (b) the herbage consumed as a proportion of herbage accumulation, i.e. the change in herbage mass between successive measurements, over the same time interval.

(iv) Efficiency of consumption, or consumption efficiency. This is the efficiency of the conversion of eaten herbage into animal product (Section 7.5).

(v) Efficiency of support energy use. This is the efficiency of usage of energy derived from non-renewable, mainly fossil fuel sources (Section 7.5).

1.4
Stability

Grasslands are plant communities which are maintained by grazing animals and management. Management may include fertilizing and sowing of new species, or the avoidance of interference. The stability of a grassland system is indicated by the amount of variation experienced by a community of both plants and grazing animals around their dynamic equilibrium following disturbance (Fig. 1.4c). This is compatible with Margaleff's (1969) view, that 'a system is stable if, when changed from a steady state, it develops forces that tend to restore it to its original condition'. Factors such as climate, topography and selectivity of grazing interact all the time to cause rapid changes in plant and animal performance. For example, seed set and the soil seed bank vary seasonally (Section 2.3). These variations maintain plant populations within a range except in extreme situations of flood, drought or overgrazing. Stability, defined thus as resilience or homeostasis, is measured either by the variation in productivity about its long-term mean or by the time taken for the community to return to its equilibrium state following a catastrophe (Fig. 1.4c). This differs from the population ecologist's definition of stability as constancy in population size or the physicist's definition, as resistance to displacement (Preston, 1969).

1.5
Sustainability

Vegetation in any ecosystem changes gradually as one population succeeds another. This is what Tansley (1920) defined as succession. The gradual change may be progressive and associated with increased structural complexity and species diversity or it may be regressive and involve the loss of species. Succession ends, or more correctly pauses, with the creation of a 'climax' community within which plant reproduction and environmental factors are relatively balanced so that the species composition and productivity fluctuate in response to seasonal changes in weather and other factors, without long-term trend. This climax community is an equilibrium composition of species.

Sustainability is a measure of the degree of difficulty which management encounters in maintaining a community (Fig. 1.4d). An insect invasion illustrates how grasslands respond to a perturbation (Fig. 1.4d). In Australia, the introduction of aphids (spotted alfalfa aphid, *Therioaphis trifolii*, and bluegreen aphid, *Acyrthosiphon kondoi*) caused substantial losses in productivity of lucerne (*Medicago sativa*) in 1977–79 but little long-term change in yield, as shown in the left-hand figure. By contrast, introduction of the fungal pathogen anthracnose (*Colletotrichum gloeosporioides*) caused degeneration of Townsville stylo (*Stylosanthes humilis*), the single legume species in northern Australia, to the extent that the pasture species was discarded within five years, as shown schematically in the right-hand figure. Grasslands have varying degrees of sustainability in response to stress (Fig. 1.4d). Agronomic manipulation may enable us to sustain grassland systems which might otherwise degrade: e.g. legume-based grassland degradation associated with soil acidification is combated using tolerant rhizobia, liming and gypsum. Unsustainable systems are exemplified by the widespread desertification which has occurred in parts of Africa and Australia due to the stresses of overgrazing and salinity.

Appreciation of the dynamic nature of climax grassland communities will cause agronomists to be relatively unconcerned by seasonal or short-term perturbations in population structure and productivity. The dynamics of species composition and productivity are discussed in Section 3.5.

1.6
Animal productivity

The types of domestic animals and their contribution to man are reviewed elsewhere (e.g. Spedding, 1971; Nestel, 1984). Briefly, cattle account for approximately 70 per cent of the world's domestic animal units when livestock of various classes are compared using weightings for size (FAO, 1979). Sheep and goats make up 11 per cent of the world's animal units and buffaloes 10 per cent. Most of these animals are in developing countries (Table 1.3). The distribution of livestock reflects:

(i) the abundance of grasslands – at least three countries which are major producers of any class of livestock have extensive natural grasslands;

(ii) the distribution of cropping systems which

require livestock, e.g. water buffalo, and which generate crop residues for animal feed; and
(iii) the distribution of the human population, which accounts partially for the importance of India and China as livestock producers (Fig. 1.6).

The densities of the animal populations reflect not only grassland productivity but also management. Average stocking rates on permanent grassland range from 0.1 animal unit per ha in Australia to 1.5 animal units per ha in Europe and 9 animal units per ha in parts of Asia. Of course such averages mask the variation in stocking rates which occurs within each country. In the great grasslands the biomass of herbivores in Africa ranges from 30 to 100 kg per ha and in northern Australia it ranges from about 10 to 60 kg per ha, including native herbivores such as kangaroos, which probably amount to only 1 kg per ha (Mott, Otthill & Weston, 1981).

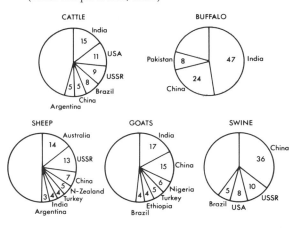

Fig. 1.6. Distribution of livestock among countries. Values are percentages of each of the classes of livestock, for which Table 1.3 gives the total numbers. (From Temple & Reh, 1984.)

Table 1.3. *The distribution of world grasslands and livestock*

	Pastures		Number of animals ($\times 10^6$)				Animal units[b]
	10^6 ha	%[a]	Cattle	Buffaloes	Sheep	Goats	
Developed market economies							
North America	265	14	137	0.2	13	1	111
Western Europe	72	19	101	—	92	10	91
Oceania	496	63	41	—	194	—	52
Other	83	52	17	—	31	6	17
Total/mean	916	29	296	0.2	330	17	271
Developing market economies							
Africa	688	30	131	—	109	112	127
Latin America	524	26	270	0.2	112	29	230
Near East	190	16	47	4	145	66	63
Far East	33	4	254	93	67	113	314
Other	0.6	0.7	0.6	—	—	—	0
Total/mean	1436	22	703	97	433	320	734
Centrally planned economies							
Asia	344	30	71	33	90	67	105
Eastern Europe and USSR	389	17	144	0.7	174	7	134
Total/mean	733	21	215	34	264	74	239
All countries	3085	23	1214	131	1027	411	1244

[a] Percentage of total land area which is devoted to permanent grassland.
[b] Animal units are total numbers of cattle, buffalo, sheep and goats equated on a relativity of 0.8, 1, 0.1 and 0.1 respectively.
Source: FAO (1979).

Livestock productivity may be assessed per hectare or per animal; this is discussed in detail in Section 7.2. Aggregates suggest that meat productivity of beef cattle and buffaloes ranges from 20 kg per animal unit per year in developing countries to 96 kg per animal unit per year in developed countries. Likewise, milk productivity ranges from 670 to 3100 kg milk per animal unit per year (Henzell, 1983). This variation arises in part because animals provide most of the non-human power used in agriculture in developing countries: China alone has over 30 million buffaloes (Table 1.3). The productivity of sheep and goats is 40–65 kg meat per animal unit per year.

1.7
Grasslands and grassland systems within farming systems

At the beginning of this chapter we pointed out that grassland agronomy is a science which operates within the farming system. Consequently this book, although mainly devoted to biological aspects of grassland agronomy, does conclude with a systems perspective of some of the economic and social issues which are pertinent to grassland agronomy (Chapter 8).

In describing the biological aspects of the system our bias is towards assessing productivity, efficiency, stability and sustainability, but particularly productivity and efficiency. This unequal emphasis usually leads technologists to the conclusion that grassland agronomy should become more productive: that there are large areas of native grasslands, scrub or forest which could be 'improved' and that productivity could, and should, be increased in grassland systems which are currently based on sown pastures. For example, it has been calculated that there are 300 million ha of improvable grasslands in both humid temperate regions and the wet-and-dry tropics of South America (Bula *et al.*, 1977; Cochrane, 1975 cited by Norman, Pearson & Searle, 1984) and we might

estimate that there are a further 100 million ha in Africa so that the total area of practically improvable grassland, if we include Asia and Australia, probably exceeds 700 million ha. Much of this 700 million ha is in the belt of highest NPP (Table 1.2).

The biological advantages of intensification of productivity are well documented (Henzell, 1983). However, in Chapter 8 we ask whether intensification is necessary technically or appropriate socially. For example, in tropical South America, about one-third of the total land area is classified as having a high likelihood of erosion, 40–50 per cent has a low effective cation exchange capacity and almost all (82–96 per cent) is classed as deficient in the major elements nitrogen and phosphorus (Norman *et al.*, 1984). Intensification of grasslands on such soils is better than intensification of cropping (Section 8.3), but it is still not necessarily 'improvement'. Agronomists now have an awareness of technology appropriate for low-input and unstable agriculture (Sanchez & Salinas, 1981). Indeed, we now recognize that, on average, 'spelling' (removal of stock) may increase the NPP for periods of at least seven years in rangelands (Lacey & van Poollen, 1981). Conversely, efforts to intensify grassland systems have failed when there has been no appreciation of the constraints of low soil fertility and rainfall, and of the biological and economic sustainability of traditional husbandry. Breman & Wit (1983) have reviewed such problems, and suggest strategies of management within traditional livestock farming systems for the Sahel, the semi-arid grassland region bordering the Sahara. The most hopeful scenario is that developing countries will follow the trend which is occurring in grassland production in developed countries: highly managed intensification where it is economically and biologically feasible to maintain stability and extensification, including reversion of grasslands to woodland and forest, for unstable or unsustainable grassland systems.

2

Generation

Seed constitutes both the start and the end of the life cycle of most plants. This life cycle involves seedling generation, survival and growth, flowering and further seed formation and the addition of this seed to the soil (Fig. 1.2).

Structural and morphological differences exist among seeds (Fig. 2.1), seedlings (Fig. 3.2) and mature plants. In this chapter we are concerned mainly with the seeds from plants of the families Poaceae (monocotyledonous, grasses) and Fabaceae (dicotyledonous, legumes). These families are the most common in grasslands; they occur in natural grasslands and they have been actively collected and selected for sowing. Unlike the case for species selected as food crops from these same families, seed size has not been a major criterion for selection of herbage plants and thus the majority of herbage species in use today have small seeds which weigh less than 2 mg (Table 2.1). Some fodder and dual-purpose species are exceptions.

In this chapter we consider first the sources of seed and the means by which this seed enters the soil and then describe the fate of this seed by following a conceptual model (Fig. 2.2) in which seed can be lost through various pathways or filters as it progresses through germination. We close the chapter with a discussion of vegetative generation, which is common in most grasslands and virtually the only method of establishment or maintenance of some (e.g. sterile) cultivars.

2.1
Sources of seed
A soil seed bank is the reserve of viable seed present in the soil and on its surface. There are three sources from which seed enters the bank (Fig. 2.2), namely (i) parent plants *in situ*, a direct source, (ii)

parent plants displaced in time or space so that the seed reaches the soil seed bank by a range of dispersal mechanisms and (iii) sowing by man to meet particular objectives, even though the natural sources also apply to these 'sown' pastures after they have been established.

Production of seed by herbage plants depends, among other factors, on the density of sites for flowering and subsequent seed formation. Site density is determined by the relationship between plant number per unit area and individual plant size because seed production is not simply a product of the density of parent plants, as discussed in Chapter 4. Seed number per fertile shoot is variable, as is subsequent production (seed number) per unit area; Krylova (1979) reviews examples of naturalized herbage legumes in the USSR in which seed numbers of individual species rarely exceed 150 per m^2 (Table 2.2). Total annual seed production or 'rain' per unit area from all species may be much higher; in tall-grass prairies in North America it reaches 20 000 seeds per m^2 (Rabinowitz & Rapp, 1980).

Once seed is produced and separated from the parent plant it must survive, enter the soil seed bank and go on to become a reproducing individual. Herbage plants and weeds with which they compete have developed many strategies to enable this cycle to proceed (Crawley, 1983).

The diversity of strategies and environments makes it difficult to generalize about the role of the parent plant in the survival and success of the seed. The amount of seed produced by a parent plant is not in itself a measure of the fitness (in the Darwinian sense) of a line. For example, the survival rate of prolific-seeding herbage species in comparison to that of low seed-yielding species may be: (i) lower when there is severe competition between species, (ii) lower where there is density-dependent seed predation, or (iii) higher when it is necessary to satiate predators such as ants for successful establishment.

The movement of seed from a parent plant to the seed bank depends on (i) the pattern of release of seed from the parent plant and (ii) dispersal mechanisms. Grasses and legumes have been classified on the basis of periods of seed release (Dudar & Machado, 1981). Seasonal patterns of release, as distinct from production, are illustrated in Fig. 2.3: release may be sudden, as in *Poa trivialis*, or delayed as in *Holcus lanatus*. Once released, seed is dispersed so as to reduce the likelihood of competition with the parent plant and other, non-dispersed, emerging seedlings. Dispersal may also transfer seed to areas with fewer specialist predators (Harper 1977; Crawley, 1983).

There are three dispersal mechanisms:
(i) dispersal by natural elements or by adhesion or attachment to the coats of animals;
(ii) ingestion by grazing herbivores with subsequent passage through the digestive tract and defecation of herbage seed in new sites; and
(iii) dispersal by non-herbivorous predators such as ants.

The significance of natural elements in seed dispersal depends on (i) the species (ii) the climate, and particularly the weather at the time of release of seed from the parent plant, and (iii) the topography of the site. Morphological differences between species affect their seed dispersal. For example, the whole inflorescence of *Chloris truncata* (windmill grass) dehisces and is transported readily by wind whereas in *Lolium* spp. (ryegrasses) the inflorescence stays attached firmly to the parent plant and wind dispersal, even of individual seed, is relatively unimportant. Dispersal by wind of legume seed is negligible as the seed is dense and usually without appendages such as glumes or bracts.

Fig. 2.1. The main anatomical and morphological features of a typical grass seed (top left), embryo (top right) and legume seed. The grass seed is botanically a fruit, or caryopsis.

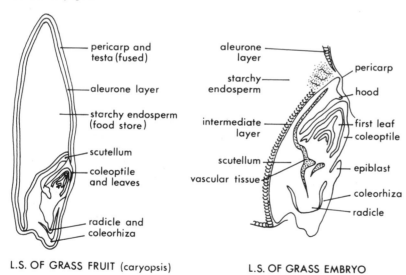

pericarp and testa (fused)
aleurone layer
starchy endosperm (food store)
scutellum
coleoptile and leaves
radicle and coleorhiza

aleurone layer
starchy endosperm
intermediate layer
scutellum
vascular tissue
pericarp
hood
first leaf
coleoptile
epiblast
coleorhiza
radicle

L.S. OF GRASS FRUIT (caryopsis)   L.S. OF GRASS EMBRYO

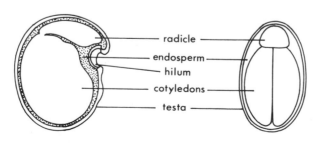

radicle
endosperm
hilum
cotyledons
testa

L.S. OF LEGUME SEED

Table 2.1. *Seed characteristics which form the basis of a sowing guide: minimum germination and pure seed percentages as per legislation in New South Wales, Australia, 1984*

Species	Germination[a] (%)	Pure seed (% by mass)	Average seed weight (Thousands/kg)	Usual sowing rate (kg/ha)
Legumes				
Lucerne (*Medicago sativa*)	60	98	440	2–15
Red clover (*Trifolium pratense*)	60	97	550	2–8
Cowpea (*Vigna unguiculata*)	70	97	4–15	10–35
Joint vetch (*Aeschenomene falcata*)	50	95	375	1–3
Kenya white clover (*Trifolium semipilosum*)	50	97	850	1–3
Strawberry clover (*Trifolium fragiferum*)	60	97	820	1–2
Subterranean clover (*Trifolium subterraneum*)	70	98	120	2–8
White clover (*Trifolium repens*)	60	97	1600	1–4
Woolly pod vetch (*Vicia villosa* ssp. *dasycarpa*)	50	99	26	5–10
Grasses				
Cocksfoot (*Dactylis glomerata* (Mediterranean))	65	85	1800	2–5
Phalaris (*Phalaris aquatica*)	65	97	880	2–6
Fescue (*Festuca arundinacea*)	65	95	420	4–6
Perennial ryegrass (*Lolium perenne*)	65	97	530	5–20
Kikuyu (*Pennisetum clandestinum*)	60	95	400	1–6
Rhodes grass (*Chloris gayana*) (tetraploid)	25	40	4000	1–6
Setaria (*Setaria sphacelata*)	20	60	1300–1900	2–5

[a] This figure does not include hard seed.

Table 2.2. *Seed productivity and seed yield of legumes in different natural zones of the USSR*

Species	Seeds/fertile shoot[a]	Seeds/m^2	Species	Seeds/fertile shoot[a]	Seeds/m^2
Forest zone			Subalpine meadows		
Lathyrus pratensis	10–17	52–96	*Lotus caucasicus*	0–29	0–110
Trifolium hybridum	31–80	463–3837	*Medicago falcata*	16	34
Trifolium pratense	2–12	4–78	*Onobrychis biebensteinii*	0–14	3–85
Vicia cracca	4–91	3–35	*Orobus cyaneus*	0–21	0–4
			Trifolium ambiguum	82	31
Forest steppe zone			*Trifolium pratense*	0–50	0–103
Medicago falcata	1799	—	*Trifolium trichocephalum*	0–5	0–2
Onobrychis arenaria	27–172	—	*Vicia truncatula*	0–5	—
Trifolium alpestre	24	37	*Vicia variabilis*	0–7	—
Trifolium montanum	87	155			
Trifolium pratense	144	—			

[a] Ranges represent year-to-year fluctuations; single values are averages.
Source: Krylova (1979).

Seed dispersal does not stop when seed comes into contact with the soil; dispersal continues with run-on and run-off of water, and movement within the soil. In all grasslands there is variation in the topography and this causes variation in wind and water dispersal and creates different microsites for the lodgement of seed. The importance of these factors was assessed in a study by Torssell (1973) of a grazed annual legume (*Stylosanthes humilis*) and perennial grass. High intensity rainfall resulted in concentrations of seedlings to an average of 2000 legume plants per m² in depressions where a 50:50 grass–legume mixture became established; this concentration was due to seed wash from surrounding bare slopes where only two grass plants per m² had become established.

The potential for domesticated livestock to disperse seeds over wide areas has been reviewed by Suckling & Charlton (1978). Legume (clover and medic) burrs readily attach to the fleece of sheep and this is a major means of dispersal of these species. In the demographic model of *Stylosanthes hamata* (Fig. 2.2) cattle ate up to 450 kg of pods per ha in one season with maximum daily intakes of about 500 g per head. When penned, cattle voided about 21 per cent of the ingested seed. A similar percentage of kikuyu (*Pennisetum clandestinum*) seed was voided in another experiment; Humphreys (1981) reviews these examples.

The percentage recovery of seed in the faeces of grazing herbivores varies with (i) the herbage species and (ii) the type of livestock. Neto & Jones (1983),

Fig. 2.2. Models to partially describe pasture generation dynamics from seed and vegetative propagules. (a) General model; (b) demographic model for grazed tropical legume *Stylosanthes hamata* in northern Australia. The models trace the fate of seed in the soil seed bank (left-hand side) and seed or propagules produced from plants (right-hand side). Seed moves from the right to the left via dispersal mechanisms and it moves from the left back to the right to become new plants. The figures give the number of seeds per m². The plus signs represent net additions to pools between December and March. ((b) adapted from Gardener, 1981.)

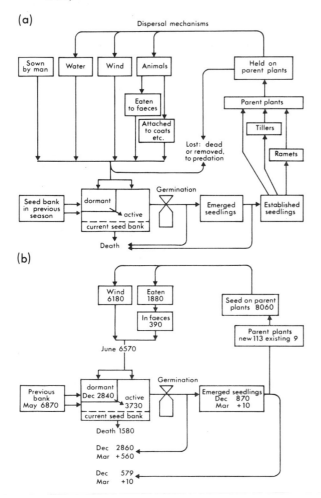

Fig. 2.3. Time course of the release of seeds of four species of grass in an English meadow. (a) ▲, *Festuca rubra*; ●, *Holcus lanatus*. (b) ●, *Poa trivialis*; ▲, *Agrostis stolonifera*. (From Mortimer, 1974.)

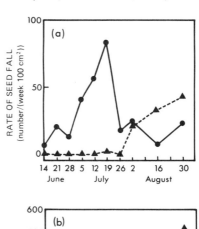

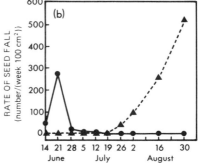

testing a wide range of tropical and subtropical legumes and grasses recovered 51, 20 and 11 per cent of ingested seed from cattle, goats and sheep respectively. A maximum recovery of 36 per cent of seed ingested was obtained for *Trifolium semipilosum*. Seed recovered from cattle faeces is generally of low viability but cattle dung is likely to create most space for germination of seedlings. The variability in emergence of voided seed of herbage species from among different livestock (cattle, antelope and rabbit) reflect real differences in seed recovery related to both type of pasture and type of animal as well as differences due to selectivity of the diet of the grazing herbivore (Wicklow & Zak, 1983).

Seed-harvesting ants are an example of non-herbivorous predators that cause some seed dispersal; these are considered in detail in Section 2.3.

2.2
Sown seed

Establishment which commences with sowing is regarded as the most hazardous phase in grassland management. The success of sowings is related directly to the extent to which competition from existing vegetation and potential vegetation (from seed already in the soil seed bank) is reduced so as to create biological space that can be exploited by sown seed of known quality. Traditionally, man has attempted to adapt the sowing environment to achieve establishment rather than select varieties which are easy to establish (Silcock, 1980).

Most seed is sown at the rate of 100–170 seeds per m^2 (Table 2.1). However, natural seed loads already in the soil commonly range from 5000 to 50 000 seeds per m^2. Sown seed must therefore compete for germination microsites the properties of which will differ from species to species and often between individual seed lots of the same cultivar (Naylor, Marshall & Matthews, 1983). These sites must provide all that is needed for germination (correct temperature, water, light and oxygen (see Section 2.4)), leaf expansion (temperature, light, vapour pressure deficit and exposure (see Chapter 3)) and root penetration (the right soil particle size, pore distribution, aeration, pH, mineral element availability and soil moisture content (see Chapter 3)).

Good quality seed is important in order to take full advantage of the investment put into the improvement of grasslands by the sowing of seed of superior genotypes. Seed certified under OECD or other legislated standards ensures varietal purity and minimum levels of quality. These quality standards vary between species (Table 2.1). Seed quality and vigour vary with

the site of production and with storage conditions. Quality may be determined by purity, germination percentage, dormancy and hardseededness characteristics and percentage of perfect, vigorous seedlings (ISTA, 1976). Measurement of these characteristics enables calculation of the percentage of pure live seed within a lot and hence allows an accurate determination of sowing rate, thus:

$$R + E \times 1/S \times 1/P \times 1/G \times 1/Q \qquad (2.1)$$

where R is the sowing rate (kg per ha of seed including impurities), E is the desired population (thousands of plants established per ha), S is the seed count (thousands per kg; Table 2.1), P is the proportion of pure seed (by mass, Table 2.1), G is the proportion of seeds which germinate (Table 2.1) and Q is the seed quality (proportion of germinated seeds which are perfect and vigorous).

The principle of sowing is to give the sown species the greatest competitive advantage over existing plants (if present) and plants which may germinate from the soil seed bank. If there are no competing plants or seeds, then the time and method of sowing simply optimize the time, rate and depth of sowing, the arrangement of the plants and the structure of the soil (through cultivation and seedling nutrition including seed treatment).

2.2.1
Time, rate and depth of sowing

These managerial factors take account of the adaptation of the cultivars selected to growth conditions, particularly the optimum temperature for their germination, emergence and early growth. They together determine the availability of soil water to the seed and the ambient temperature. Deep soil (say 2–5 cm) has the advantage of usually having more available water than does topsoil and a smaller amplitude of daily fluctuations in temperature.

The interaction between these factors can be important in dry temperate or tropical climates where temperatures at the soil surface can exceed 50°C at midday and thereby result in seed death due to high temperatures, lack of water, or both. In practice, maximum sowing depth is constrained by seed endosperm reserves, which differ markedly between species (see seed weights, Table 2.1). These reserves must be adequate to support elongation of the hypocotyl or epicotyl until seedlings emerge above the soil surface and begin photosynthesizing. The rate of elongation in turn depends on the genotype, temperature and amount of soil water. One of the most common causes of failure in establishment of small-seeded species is sowing too deeply.

2.2.2

Method of cultivation and sowing

Cultivation affects the size distribution and packing of soil particles. This determines the impedance of the soil to elongation of the radicle and hypocotyl or epicotyl and also the rate of water movement to the seed. Perry (1976) regarded aggregated soil with a mean particle size 0.1–0.2 of the diameter of the seed to be the ideal growing medium. In addition to providing tilth adjacent to the seed, the aim of cultivation is to optimize the condition of the soil surface to (i) minimize upward impedance, i.e. as encountered by the seedling just prior to emergence, (ii) minimize downward impedance (for water infiltration and to reduce run-off) and (iii) maintain gas diffusion through the soil to avoid short-term anaerobiosis through crusting and slaking of the surface.

Methods of cultivation (or lack thereof) and subsequent sowing strategy can be classified into three groups: (i) surface sowing, i.e. without soil disturbance; (ii) direct drilling, where seed is sown with one pass of the machinery, resulting in minimal disturbance of the soil or (iii) sowing in prepared seed beds, which involves a variable number of cultivations. The advantages and disadvantages of each of these methods are summarized in Table 2.3.

Table 2.3. *Management implications of the three main methods of sowing*

Characteristic	Surface sowing (No soil disturbance)	Direct drilling (One pass with minimum soil disturbance)	Sowing in prepared seed bed (Variable number of cultivations)
Risk of erosion	Low	Minimal	High
Fuel usage	Low	Low	High
Labour costs	Low	Medium	High
Implements used	Few, e.g. by hand or spreader, or specialist, e.g. aircraft	Tractor and sower	Variable number: ploughs, cultivators, harrows, sower
Surface trash	Not removed; minimized by grazing, burning or perhaps herbicides	Removed by grazing and herbicides	Incorporated
Root competition	High	High, reduced by use of herbicide	Low
Light competition	Variable, depending on management	Variable, depending on management	Low with good weed control
Soil seed contact	Poor; can treat seed	Variable, depending on management	Good
Insect/pest control	Poor	Poor; may need pesticide	Best
Sowing depth	No control	Control depends on machinery, soil moisture and surface trash	Good control
Time to grazing	Up to 2 years	Grazing in year of sowing	Grazing in year of sowing
Soil structural problems	No effect	Minimal; enhances structure in long term	Can be created; can use machinery to break subsoil compaction
Soil moisture	Critical	Critical; conserved under killed sward	Not as critical
Soil temperature	Seed exposed to extremes	Soil temperatures may be reduced. Late sowing can be a problem	Depends on soil type and depth of sowing
Fertilizer placement	On surface, poor control	Can place below or beside seed; good control possible	Can place below or beside seed; good control possible
Opportunity cost	Variable	Low; can graze to sowing	High

Surface-sown seed is dispersed by hand, ground machinery or aerially into existing grassland swards which may be alive, dead (e.g. after they have completed their life cycle) or killed, or onto the soil surface where there is no vegetation as e.g. after fire. Surface sowing is generally less reliable than sowings involving soil disturbance although costs and the risk of soil erosion are generally lower. Aerial seeding is successful in non-arable hill country in New Zealand (Suckling, 1976), the United States (Peters, 1976), Iran (Campbell, 1979) and southern Australia (Campbell, 1982). Conditions of temperature and moisture for seed germination are more severe at the soil surface than in the subsoil. Evaporation from the seed coat may exceed the rate at which the seed absorbs moisture and the level of seed moisture which is critical for germination may not be reached. Seed on the soil surface can imbibe water only through that area of the seed which is in contact with the soil surface; this puts large seeds with greater surface area at a disadvantage in comparison to small-seeded species. Water uptake may be enhanced by the use of seed coatings such as lime (McWilliam & Dowling, 1970).

The practice of direct drilling of seed into previously undisturbed soil causes minimal soil disturbance (Table 2.3). This is frequently referred to as sod-seeding. It is usually preceded by grazing, cutting or killing existing vegetation and followed by management to maximize emergence and establishment. The subject is well reviewed (Swain, 1976; Decker & Dudley, 1976; Naylor, Marshall & Matthews, 1983). Machinery used for direct drilling is intended to perform a number of functions: (i) tilling of a narrow band of soil along the line of a furrow; (ii) opening of a furrow; (iii) metering of the seed to control sowing density; (iv) placement of seed in contact with the soil; (v) covering of the seed and (vi) consolidation of any tilled band. Machinery and techniques, and consequently the achievement of these aims, vary in practice.

Naylor *et al.* (1983) list ten guidelines for successful sowing by direct drilling:
 (i) manage the old sward to reduce sward height and minimize the standing biomass;
 (ii) choose a suitable time of year;
 (iii) kill the old sward with a suitable herbicide;
 (iv) wait until the herbicide does not affect sown seed;
 (v) reduce the amount of trash consistent with the need to conserve moisture and prevent erosion;
 (vi) if possible select seed of high vigour;
 (vii) sow using a machine able to overcome problems of a particular site, e.g. tussocks, a matted sward, soil texture;
(viii) ensure that seed is covered and the drill slits consolidated;
 (ix) incorporate crop protection chemicals as needed; and
 (x) manage the young sward to ensure survival of established plants.

Prepared seed beds are the result of one or more cultivations prior to sowing (Table 2.3). The main aim of preparation is to minimize competition from either existing plants or potential plants in the form of seed. This is done by cutting roots, incorporating plants into the soil and reducing the soil seed pool by promoting early germination; the seedlings are then removed by further cultivation. A major, although often neglected, advantage of prepared seed beds is the reduction in root competition which is usually of greater importance than light competition in determining the success of seedling establishment (Silcock, 1980). Care must be taken not to tie up nitrogen by incorporating large amounts of organic matter and cultivation should not be excessive, in order to avoid problems with soil structure.

The key criteria determining the selection of a sowing method are: cost, timeliness, the availability of labour and machinery, the amount of competition and the purpose of sowing, and the time when herbage is needed (Table 2.3). Similar populations of desired grassland species can be achieved via different sowing methods (Table 2.4) but at varying cost and over varying time periods.

2.2.3
Nutrition and seed treatment
Fertilizing or seed treatment may be desirable in order to favour the sown species and correct mineral deficiencies. Fertilizer is best placed below or beside the sown seed. Coating the seed with nutrients (e.g. phosphorus, sulphur and molybdenum) may enhance establishment and reduce fertilizer costs. Where appropriate, seed should be treated with an inoculum of *Rhizobium* (for legumes to be sown in soil not having native or previously introduced rhizobia), fungicide and pesticide (where ants and other predators may reduce the sown seed bank, see e.g. Dowling & Linscott (1983), Section 2.3).

The above aspects of sowing have been considered largely with respect to species sown by man in the absence of any other species. In practice, however, the sowing strategies chosen are a compromise between what may be optimal for the sown species, i.e. what would be most likely to achieve the desired population, as shown in Eqn (2.1), and what is desirable for diminishing the competition from other species. For example, the best time of sowing then becomes that time when the temperature and amount of water may not be optimal for the sown species, but are increasingly satisfactory for the sown species and increasingly unsatisfactory for existing species. The optimum time of sowing a tropical or subtropical C_4-grass species (having the Hatch-Slack photosynthetic pathway and a temperature optimum for germination of, say, 30°C) into soil containing seed of mostly C_3-species (which have Calvin cycle photosynthesis and a temperature optimum of 20–25°C) could be either spring or early summer in, say, a subtropical environment. The best time of sowing, however, is clearly late spring or summer when temperature and moisture conditions combine to the benefit of the C_4-species and disadvantage of the C_3-species.

Three further methods are used to achieve management compromises between sown and existing species. These are grazing, herbicides and burning.

2.2.4
Grazing

Using animals to reduce standing herbage is the least costly way of minimizing existing competition prior to sowing. Together with burning, it is also the least effective (Cook, 1980). Reduction in competition depends on grazing pressure, class of animal (e.g. high stocking rates of non-lactating animals are preferable), pre-sowing grazing management (stock may have to be introduced up to six months prior to sowing) and climate, particularly short- and long-term variation in rainfall. In areas of reliable rainfall grazing prior to surface sowing improves establishment. In

areas of unreliable rainfall retention of ungrazed material improves establishment of surface-sown clover, lucerne and ryegrass (Table 2.5) because of greater moisture retention and greater atmospheric humidity close to the unburied seed. In other situations large amounts of residual vegetation may inhibit establishment due to physical or chemical factors.

Grazing may also be used where there is a need to remove material which physically restricts sowing or cultivation machinery. The trampling effect of grazing animals may improve establishment in instances of surface sowing or vegetative propagation. In an Israeli example, soil surface disturbance of dry range by sheep compensated for increased seed production in the non-grazed area so that seedling regeneration was comparable in grazed and non-grazed areas (Crawley, 1983, p. 81). The transfer of seed in dung has been discussed previously.

Slashing or mowing is a costly alternative to grazing and the material remaining may obstruct sowing or cultivating machinery and reduce the efficiency of a contact herbicide. Forage harvesting removes the material; the efficiency of silage and hay production are considered briefly in Section 7.3.

2.2.5
Herbicides

A herbicide will usually be required to kill or suppress competing plants, particularly when surface sowing or direct drilling is being carried out. The choice and effectiveness of a herbicide or herbicide mixture will depend on the species to be controlled,

Table 2.4. *The results of various techniques used to sow lucerne* (Medicago sativa) *at 8 kg/ha into a perennial ryegrass* (L. perenne) *and subterranean clover* (Trifolium subterraneum) *pasture in spring in Canberra, Australia*

	Relative cost	Density (plants/m^2)	Dry matter (kg/ha) in year: 1	2	3	Total dry matter (kg/ha)
No herbicide						
Low cost sod-seeding	1.0	0.1	4	240	4	250
Minimum surface tillage	1.6	16	400	1900	1400	3700
Shallow mould-board ploughing	3.0	38	1200	2100	1600	4900
With paraquat						
Low cost sod-seeding	2.6	12	400	2400	1100	3900
Minimum surface tillage	3.2	29	600	2200	1000	3800
Shallow mould-board ploughing	4.6	42	1300	1700	1400	4400
With dalapon						
Low cost sod-seeding	2.3	10	350	2600	1300	4250
Minimum surface tillage	2.9	29	1000	2200	1200	4400
Shallow mould-board ploughing	4.3	39	1300	1700	1400	4400

the species to be retained, their stage of growth, the time required from spraying to sowing and the relative costs of the materials. Two groups of herbicides are commonly used before sowing and both are inactivated on contact with soil: these are contact and translocated herbicides. Contact herbicides, of which the bipyridyl herbicides (e.g. paraquat) are the most common, are most effective against young annual weeds or for short-term suppression of perennial grasses (e.g. Decker & Dudley, 1976). Paraquat-treated herbage is rapidly desiccated and seed can be sown 2–3 days after spraying. Translocated herbicides are typified by glyphosate, the subject of a recent extensive review (Grossbard & Atkinson, 1985). Translocated herbicides are most effective against advanced weeds and for killing, or at low rates of application, suppressing perennials. Translocated herbicides require a greater time from spraying to sowing than do contact herbicides. Naylor *et al.* (1983) review this subject in greater detail.

Natural products of microbial breakdown of the residues from previous crops can affect germination. They include acetic acid, tannins and phenolic compounds causing phytotoxicity. In one Welsh study, use of both paraquat and glyphosate followed by direct drilling inhibited germination and establishment when the interval between spraying and drilling was not less than 21 days whereas burning or removal of old sward improved establishment (Davies & Davies, 1981).

2.2.6
Burning
The use of fire can enhance establishment usually by reducing competition for light by retarding the growth rate and removing the foliage of the competing plants. The extent of plant death depends on the temperature of the fire and the sensitivity of the species to burning. In northern Australia species such as *Sorghum plumosum*, *Themeda australis* and *Sporobolus elongatus* are particularly sensitive whereas the widespread black spear grass (*Heteropogon contortus*) is resistant to burning (Cook, 1980). The spear grass is therefore able to compete, even following burning, with surface-sown or direct-drilled species.

An understanding of the adaptations of grassland species to fire will enable better use of burning as a management tool for establishment. The adaptations include the ability to regenerate from adventitious buds, heat-promoted release of seeds from fruits, heat-promoted germination of hard seeds (Section 2.3), long-lived soil seed banks and fire-related reproductive cycles (Hodgkinson *et al.*, 1984). Gill (1981) recognizes two groups of species based on response to fire, namely sprouters, where plants can survive all leaves being scorched, and non-sprouters, where plants die if all leaves are scorched and regeneration must be from seed.

2.3
Dynamics of the seed bank
The size of a soil seed bank reflects inputs from current and past generations of parent plants, seed arriving by dispersal pathways from other parent plant communities and sown seed, and losses from predation, death or germination (Fig. 2.2). Thus in examining the population dynamics of any grassland species it is necessary to measure quantity, viability and duration of stay of seed of that species in the seed bank. Generally, the composition of seed banks never reflects the current composition of the grassland and the banks may contain seed of pioneer species which disappeared early during succession. White clover is one pasture plant which has proportionally more seed in the seed bank than there are plants in the pasture. Seed densities are usually 200–300 per m^2 but may

Table 2.5. *Effect of grazing management[a] and herbicide treatment on the success (per cent) of establishment of surface-sown seed of four pasture species*

	Herbicide			No herbicide		
	Heavy	Light	No grazing	Heavy	Light	No grazing
Subterranean clover	6.1	12.0	34.9	7.2	16.4	29.8
Lucerne	3.5	5.6	21.5	6.9	14.6	21.1
Phalaris	0.5	1.0	2.7	0.2	1.3	7.1
Ryegrass	8.0	12.5	23.2	7.3	4.5	14.4

[a] Heavy, light and no grazing had herbage cover at sowing of 750, 1250 and 3300 kg dry matter per ha respectively.
Source: McWilliam & Dowling (1970).

reach 15 000 per m^2 (Burdon, 1983). The rate of seed decline as a percentage of total soil seed reserves is greater for grasses than for dicotyledonous species when seed inputs are stopped (Williams, 1984). Furthermore, in dense undisturbed swards, the chances of seeds of most species becoming incorporated into a permanent seed bank are low. Dormancy status and seed size affect the chance of seed burial, e.g. seeds of *Poa* spp., which are dormant and small, predominate in seed banks under English leys (Howe & Chancellor, 1983). Methods of measuring or estimating these parameters are reviewed by Roberts (1981).

2.3.1
Seasonality of seed banks

English grasslands have been grouped into four types on the basis of the seasonal release of seed from parent plants and the fluctuations of seed in their seed banks (Fig. 2.4). These are:
 (i) Species with transient seed banks during the summer (Fig. 2.4a). This group includes cocksfoot (*Dactylis glomerata*) and perennial and Italian rye-grass. These species release seed in late spring and summer and germinate in the autumn. The seed of these species is characterized by large size and/or elongated structure which includes projections such as awns, lack of pronounced after-ripening or dormancy mechanisms, the ability to germinate over a wide range of temperatures and the ability to germinate both in the light and in continuous darkness. These features facilitate rapid germination soon after seed fall and result in poor persistence of ungerminated seed in the seed bank. The latter has important management implications where these sown species are replaced by native grasses with persistent seed banks.
 (ii) Species with transient seed banks present during the winter (Fig. 2.4b). This is seen as an adaptation which delays germination until the beginning of the spring growing season in environments which have very cold winters. Imbibed seed of species in this category frequently require exposure to temperatures of 2–10°C (stratification) before they will germinate. Perennial grasses in this category may regenerate vegetatively in spring from dormant rhizomes and stolons, e.g. *Poa pratensis* and *Trifolium repens* (Grime, 1979).
 (iii) Species with persistent seed banks in which some of the seeds are at least one year old (Fig. 2.4c). In this case the majority of seed acts as in (i) with the remaining seed being incorporated into the persistent seed bank. An example is the winter-growing annual *Lolium rigidum*, a useful grassland species but also a weed of cultivation as 20–60 per

cent of ungerminated seed may persist after the autumn rains (Gramshaw & Stern, 1977*b*).
 (iv) Persistent seed banks where only a few seeds germinate in autumn (Fig. 2.4d).

Species with very small seeds fall mainly into categories (iii) and (iv) (Thompson & Grime, 1979). These are also the strategies currently being pursued in the selection of annual mediterranean legumes such as *Trifolium subterraneum*, *Vicia villosa* ssp. *dasycarpa*, *Ornithopus compressus* and *Medicago* spp. (Fig. 2.4e). These species produce seed in spring. However, as a high proportion of the seed does not imbibe water due to an impermeable seed coat or dormancy, only a proportion of that seed germinates in the following autumn. This strategy allows the species to escape moisture and temperature stress over summer, survive false or failed autumn rains and still maintain viable seed in the seed bank.

Fig. 2.4. Four types of seed banks which occur commonly in temperate regions in the northern hemisphere ((a)–(d)) and two types of seed banks postulated for legumes in mediterranean (e) and tropical pasture systems (f). □, Seeds capable of germinating immediately after removal to suitable laboratory conditions; ■, seeds viable but not capable of immediate germination. (a) Annual and perennial grasses of dry or disturbed habitats; (b) annual and perennial herbs colonizing vegetation gaps in early spring; (c) species mainly germinating in the autumn but maintaining a small persistent seed bank; (d) annual and perennial herbs and shrubs with large persistent seed banks; (e) postulated scheme for mediterranean annual legumes in the southern hemisphere; (f) postulated scheme for tropical annual legumes in the southern hemisphere. (Adapted from Thompson & Grime, 1979.)

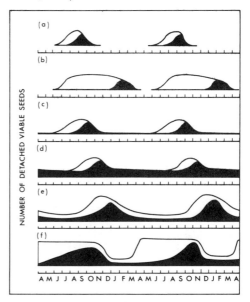

Annual tropical legumes may be seen as a variation of category (ii) species in that they must escape drought in winter or the dry season and false breaks of the wet season (Fig. 2.5f). However, in contrast to category (ii) species many of the tropical legumes form persistent seed banks (Jones, 1986).

Little is known of the seed dynamics of the tropical grasses. The native *Sorghum* spp., kangaroo grass (*Themeda australis*) and black spear grass of tropical northern Australia have seed with hygroscopic awns enabling them to bury themselves beneath the soil surface after seedfall. Seed is dormant when it is shed whereas at the break of the following wet season almost all of the seed in the seed bank is non-dormant and rapid germination occurs, mainly after the first rainstorm of 20 mm or more (Andrew & Mott, 1983). This leaves little or no seed in the soil seed bank. Tropical grasses are thus not unlike category (i) species in strategy.

2.3.2
Dormant and active seed

Seed which is alive as it leaves the parent plant may be in one of four physiological states:

(i) non-dormant, and able to germinate immediately;

(ii) innately dormant due to the presence of inhibitors which can be overcome by a seasonal stimulus like photoperiod or thermoperiod;

(iii) induced dormancy, which lasts after the stimulus that produced it has ceased to act; and

(iv) enforced dormancy, where dormancy is imposed by a factor like cold or darkness and lasts only as long as the factor acts on the seed.

The mechanisms controlling dormancy are complex, as detailed in reviews by Roberts (1972), Taylorson & Hendricks (1977) and Rolston (1978). The causes of dormancy include impermeability of the seed coat to water and gases, immaturity of the embryo, special requirements for temperature or light, the presence of inhibitors and mechanical restriction to embryo growth and development, or to radicle extension, during germination. Several of these factors may operate together (Murray, 1984). Seed may change its physiological status as it enters the soil seed bank and once it is in the seed bank, and in cases of innate, induced or enforced dormancy a change in status is a prerequisite for movement into the active seed bank (Fig. 2.2).

Three situations can be recognized where buried dormant seed banks persist for a long period of time with little movement to the active seed bank (Roberts, 1981):

(i) in soil which is periodically disturbed, seed remains in the enforced dormancy category for only a part of the time, and the mechanisms of innate and induced dormancy are responsible for keeping seed in the dormant seed bank;

(ii) in undisturbed soil, with seeds distributed fairly close to the surface (to about 20 cm) enforced dormancy is responsible; and

(iii) in undisturbed soil with seed buried deeply in uniform conditions, the intrinsic capacity for survival of seed in an imbibed state is responsible for continued dormancy.

The dormancy characteristics of seed of grassland species has considerable adaptive significance. In mediterranean and tropical monsoon climates, legume hardseededness is exploited to escape high temperature and moisture stress over summer in the former and to survive moisture and, sometimes, low temperature stress in dry ('winter') seasons in the latter. Hard seeds are unable to imbibe water and thus

Fig. 2.5. Dynamics of the seed bank of four accessions of the tropical pasture grass *Digitaria milanjiana*.
Percentage germinable (black), dormant (stipple), dead (diagonals) and apparently field-germinated (white) seed at various times after placement on the soil surface on 4 June at two sites in northern Australia: Lansdown (wet tropics, left) and Katherine (wet-and-dry tropics, right). (From Hacker *et al.*, 1984.)

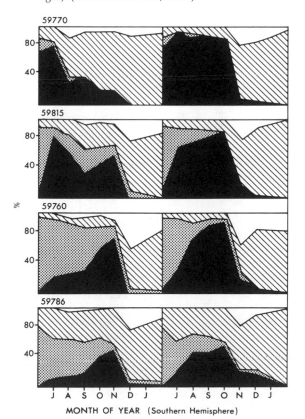

fail to germinate until hardseededness is broken. Hardseededness is thus a special case of dormancy (physical exogenous dormancy) due to water impermeability of the testa or seed coat (Fig. 2.1); it may or may not be associated with other forms of dormancy. Rolston (1978) reviews this special category of dormancy. Hard seeds have a layer of macrosclerid cells that form a palisade layer in the testa; both genetic and environmental conditions affect the proportion of impermeable seeds produced. For example, more hard seed of *Stylosanthes hamata* is produced when seed production occurs under warm conditions than in cool weather, although interactions with seed moisture content may modify this general relationship (Argel & Humphreys, 1983). Examples of the size of the dormant seed pool and the dynamics of movement from the dormant seed pool and the dynamics of movement from the dormant to the active state are considered below.

Hacker (1984) observed that grass seed dormancy has usually been regarded as a nuisance. This view is being reassessed, particularly in selecting or breeding species adapted to climates which have distinct wet and dry seasons, such as mediterranean climates and tropical savannas. Species capable of surviving extended drought, such as *Digitaria milanjiana*, require adequate seed production, seed dormancy and seed survival. The size and duration of the dormant seed bank of *D. milanjiana* varies with ecotype, site of collection of ecotypes and site of sowing (Fig. 2.5). When sown at two sites in northern Australia only the ecotypes collected in the low rainfall tropics of Africa retained dormancy through the dry season and were fully germinable at the start of the wet season at Lansdown, the site which was cooler and had more rainfall in the dry season ('winter'). By contrast, all accessions were live and non-dormant at the start of the wet season at Katherine, because of total lack of rain and high temperatures during the dry season (Hacker *et al.*, 1984).

Dormant seed enters the active seed bank naturally as a result of breakdown of, or release from, dormancy. Alternatively, seed may become active due to treatment by man before addition of sown seed to the soil seed bank. Under natural conditions a variety of pathways for release from dormancy have been recognized:

(i) In many temperate grasses exposure of moist seed to low temperatures, referred to as stratification, reduces dormancy.

(ii) In some species dormancy is maintained by the presence of the seed coat which may be the testa, pericarp or, in the case of grasses, also the glumes or lemma and palea. Dormancy may be due to a physical barrier or these organs may contain chemical inhibitors. Natural abrasion, removal of the seed coat by predators and leaching (iv below) reduce dormancy.

(iii) Impermeable hard legume seeds become permeable naturally by the action of fire, by abrasion due to wind and water movement over the ground, by microbial attack in the soil, by high and fluctuating temperatures or by passage through the digestive tracts of birds and animals.

(iv) In some temperate grasses and in the rangeland *Atriplex* species of the family Chenopodiaceae, dormancy may be broken by washing or leaching of inhibitors. In the case of *Atriplex* the 'seed' is a bladdery one-seeded fruit and the bracts which enclose the seed contain high concentrations of chloride ions which require a long period of leaching for removal under natural conditions.

Seed may be artificially treated to break hardseededness. Treatments, which usually take place prior to sowing, have been divided into 'wet' and 'dry' by Tran & Cavanagh (1984). Wet treatments include chemicals (sulphuric acid, alcohol, acetone, oxidizing agents) and extreme temperatures (hot or boiling water, liquefied gases). Dry treatments include mechanical action (manual and mechanical scarification, impaction, high pressure) and temperature (dry heating, using radiant or electromagnetic waves, field temperature fluctuations). Further aspects of regulation of seed dormancy are reviewed by Wareing (1982).

The movement of seed from the dormant to the active seed bank (Fig. 2.2) may be illustrated using two models. Firstly, a simple deterministic model describes long-term changes in plant densities of annual subterranean clover varieties (Rossiter *et al.*, 1985). Subterranean clover produces seed in spring and early summer and a proportion of this seed germinates to form the new grassland at the start of the wet autumn–winter. Our concern here is with the first part of the model (Fig. 2.6), particularly the soft (or active) seed pool at the start of the wet winter season.

The probability of the seed present at the beginning of summer surviving over the summer is P_s. This seed loses its dormancy and enters the active seed bank (softens) with different probabilities depending on whether it is seed still in the burr (the fruit of subterranean clover) or free. Seed, once free, remains free and burr seed becomes free at a rate which may depend on its age. The rate of seed softening also depends on the age of the seed. Soft seed produces established plants which give rise to adults which will in turn produce seed during the next spring. Residual dormant or hard seed from the previous season is added to current seed production to give the amount of seed in the seed bank at the beginning of the new season (Fig. 2.2).

Subterranean clover cultivars which have small seeds appear better able to survive over the summer than do cultivars with large seeds (Table 2.6). There is an approximately 20 per cent loss of seed over the summer due to death, mainly by germination of seed which fails to survive, and predation (see below). At the end of the summer about one-third to one-half of the seed produced in the previous season has softened and entered the active seed bank (Table 2.6). The proportion coming from free seed is higher than that from burr seed but free seed is rarely found until about the third year after sowing.

A second illustration of the dynamics of the seed bank is that of a grassland of annual grass and annual legume where germination occurs during the change from dry winter to monsoonal wet summer. Torssell & McKeon (1976) define three filters, namely a dormancy-breaking filter ($A_{d,s}$), a germination filter ($A_{s,g}$) and an establishment filter ($A_{g,e}$), to account for seed movement from the initial seed bank of dormant and non-dormant legume (L) and grass (G) seeds ($L_d + G_d$)

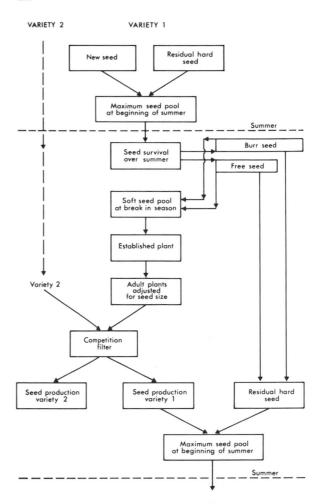

Fig. 2.6. Flow diagram illustrating a model of the changes in plant densities of annual subterranean clover, *Trifolium subterraneum*, grown as a mixture of two varieties. The diagram has been drawn in detail for variety 1: the portion for variety 2 is a mirror image. The two parts are joined through a competition filter. (Adapted from Rossiter, Maller & Pakes, 1985.)

Table 2.6. *Measurements of seed production and seed bank parameters of subterranean clover varieties used to formulate the model illustrated in Fig. 2.6*

	Dwalganup	Daliak	Yarloop	Seaton Park
Seed production				
Seeds/dm^2	190	290	45	74
Weight/seed (mg)	5.2	4.4	7.2	5.2
Proportion of seeds surviving over summer	0.84	0.87	0.73	0.83
Proportion of soft seed at end of summer from:				
Burr seed	0.37	0.36	0.53	0.57
Free seed	0.43	0.41	0.61	0.66
Proportion of juvenile plants established relative to				
soft seed numbers at end of summer	0.37	0.50	0.46	0.45
Proportion of juvenile plants surviving to adulthood				
or beginning of flowering	0.70	0.71	0.79	0.69
Prediction by the model of quantity in maximum seed				
banks (years 2–5) for each variety (Seeds/dm^2)	80–120	280–420	80	120

Source: Adapted from Rossiter *et al.* (1985).

through to the germinable seed bank ($L_s + G_s$), germinated seed bank ($L_g + G_g$) and finally to the established seedling bank ($L_e + G_e$). This model describes changes in legume and grass density from seed production through germination to establishment. The change in ratio of legume to grass (L/G) between the initial seed bank and the established seedling bank is defined as:

$$A_{d,e} = \frac{L_e/G_e \times L_e/L_d}{L_d/G_d \times G_e/G_d} \qquad (2.2)$$

where $A_{d,e}$ represents the net change over the series of filters, i.e. $A_{d,e} = A_{d,s} \times A_{s,g} \times A_{g,e}$, the product of the filter parameters. The grass *Digitaria* has no dormant seed. Thus, there is a gradual decline in the active seed bank over the dry season from May to November until the main germination event after about 18 mm of rain (Fig. 2.7). The decrease in the active seed bank is due mainly to germinations which fail ('false starts'). By contrast, the active seed bank of the legume *Stylosanthes humilis* increases by 85 per cent over the dry season. This can be explained largely by

exposure of the legume seed to extreme, fluctuating temperatures which would be expected to reduce its dormancy. The main germination event for the legume occurs at the same time as that for the grass but 20 per cent of seed remains in the dormant bank and 34 per cent of the active bank does not germinate until subsequent rainfall events. The seeds which do not germinate at the first rainfall germinate eventually but they contribute relatively little to the established seed bank due to their having to compete against already-established seedlings. In a particular experiment the germination filter for the grass had a median value of 0.76 indicating that *Digitaria* was advantaged relative to *S. humilis* at this stage. Values for the dormancy-breaking and establishment filters exceeded one, indicating advantage to the legume at these stages and overall at the site of the experiment.

The challenge is to manage the soil seed bank. Various workers suggest the desirability of this but as yet there is little sound management advice of the type proposed by Bishop, Walker & Rutherford (1983). The problem encountered by them was that of a declining gain in liveweight of grazing animals due to a declining plant density of legume (siratro). Cultivation increased the number of siratro seedlings recruited from the soil seed bank three-fold relative to uncultivated controls. Disc ploughing was better than a tyned implement due to a greater reduction in the competition from the associated grass. Thompson (1981) concluded that the amount of seed buried within the soil is correlated positively with disturbance and negatively with stress. This suggests that the seed bank is only able to contribute to production after cultivation or natural, persistent soil disturbance where there is intense selection favouring buried seed.

In some instances, grassland species are not capable of regenerating from the soil seed banks: species such as lucerne, once sown, recruit very few seedlings from the soil seed bank and sowing is necessary, at some interval depending on the longevity of the original plants, to re-establish the desired composition of the grassland.

Seed may be lost before it enters into the seed bank or directly from the soil seed bank (Fig. 2.2). These losses are due mostly to predation by ants, birds, etc. Janzen (1969) recognizes five ways in which seeds can escape predation:

(i) by being too small for predators to handle or to colonize;
(ii) by germinating too quickly to allow sufficient time for predator development;
(iii) by containing high levels of toxins;
(iv) by being produced so abundantly that predator satiation occurs; or

Fig. 2.7. Changes in the seed and seedling population in pure swards of (a) the grass *Digitaria* and (b) the legume *Stylosanthes humilis* in a wet-and-dry climate in northern Australia. Observations made in these swards were used to calculate A-values (see text) based on changes in legume to grass ratios (Eqn (2.2)). Areas on the figures are coded: I, viable seedlings; II, dead seedlings; III, dead seeds and seedlings; IV, viable germinable seeds, which is subdivisible into V, soft viable seeds and VI, hard dormant seeds.

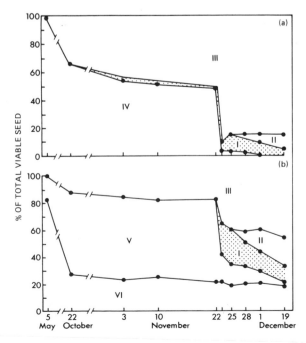

(v) by having efficient dispersal agents operating before predation: Harper (1977) and Crawley (1983) provide examples of these.

Seed losses due to predation depend on the grassland species, alternative seed sources (e.g. weed or crop seeds) and the predator. Losses may be almost total. For example, 96 per cent mortality in *Panicum maximum* (panic) in the Costa Rican lowlands (Janzen, 1969) and 90 per cent mortality in *Leucaena leucocephala* were due to predation by bruchid beetles in central America but not Puerto Rico where the beetle is absent (Janzen, 1970).

The major predators of seed are ants. Ants commonly forage for seed on the soil surface, sometimes on the parent plant and even from beneath the ground. There is a wide range of ant species harvesting a wide range of seed from many habitats (Buckley, 1982*a*). Grass seeds are often preferred over legume seeds, possibly because of their low toxicity or the ease with which they can be hulled (Buckley, 1982*b*). Ants take seed selectively: they took seed from a mixed, surface-sown Australian grassland in the ratio 1 subclover:5 lucerne:5 ryegrass:10 phalaris, indicating a preference for lighter seeds (Campbell & Swain, 1973*b*). Treatment of the seed with insecticide and to a lesser extent pelleting with lime (temperate species) or rock phosphate (tropical species) reduces losses. Campbell *santhes hamata* grasslands in northern Australia about

(1982) reviews other examples. Losses are more pronounced if temperatures favour ant activity (above 5°C) and if dry conditions follow sowing: in *Stylosanthes hamata* grasslands in northern Australia about 40 per cent of the 4000–5000 seeds per m² seed bank may be removed by ants and termites over six months of the dry season (McKeon & Mott, 1984).

Birds, mice and grazing herbivores also eat seed. Sheep consume clover burr, cattle ingest seed heads and pods.

2.4
Germination

Three conditions must be met for a seed to germinate. It must be (a) viable, (b) non-dormant and (c) exposed to a favourable environment. Once these are satisfied, germination occurs in stages:

(i) Imbibition of water by the seed, a purely physical process of uptake which occurs even in non-viable seed. Dry seed has a suction force of about -100 MPa (equivalent to 1000 atmospheres). Nearly all the water needed for germination is absorbed in 4–8 h (Fig. 2.8*a*). Later (within 12 h) the seed dehydrates to 25–35 per cent moisture (Fig. 2.8*b*) and its suction force declines to -1.5 MPa or less as germination commences.

(ii) Initiation of hormone activity, particularly of gibberellins and cytokinins, and of enzyme and respiratory activity. Metabolic aspects of germination are reviewed by Lovato (1981).

(iii) Catabolism and translocation of stored seed reserves to the embryo.

(iv) Assimilation of the now soluble reserves as an energy source for cellular activity and growth in the embryo.

Fig. 2.8. Hydration (a) and dehydration (b) curves for seeds of some pasture legumes and grasses under controlled conditions. Species are phalaris (○), annual ryegrass (△), perennial ryegrass (■), subterranean clover (●), lucerne (▲) and white clover (□). Hydration of dry seeds took place at 23°C; loss of water from fully imbibed seeds over a 12-h period took place at 20°C. (From McWilliam, Clements & Dowling, 1970.)

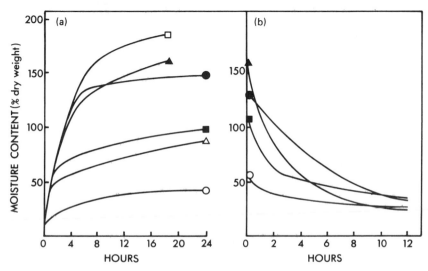

(v) Embryo growth with the first cell division occurring in the primary root tip which pierces the seed coat. This ends germination as we have defined it. Subsequent developmental stages are described in Chapter 3.

Collis-George & Melville (1975, 1978) have modelled water absorption by a swelling seed for non-limiting conditions (where seeds are immersed in water or in water-saturated soil of high hydraulic conductivity) and for limiting conditions (where e.g. seeds are sown 'dry' in anticipation of subsequent rainfall into soil with hydraulic conductivity approaching zero but perhaps above deeper soil moisture). In the water-limiting situation, water vapour transfer is possible but this may not be sufficient for all seeds to imbibe. This leads to competition among seeds for available water. Problems can be encountered because the moisture potential required for germination is above that for mould growth. Thus seed may rot before imbibition can occur. In the absence of rainfall, seed which is sown into dry soil can imbibe only if the nearby, but not contiguous, supply of water is at a potential greater than the permanent wilting potential (-15 MPa) and the rate of imbibition and germination are sufficiently rapid to minimize attack by soil-borne fungi. Imbibition results in hydration of cells leading to swelling of the embryo and endosperm which may cause the softened seed coat to break. Legumes imbibe more water more rapidly than grasses; nearly all of the water needed for germination is absorbed in the first 4–8 h (Fig. 2.8a) unless the presence of a seed pericarp slows the rate of water uptake, as in crimson clover and sainfoin (Cooper, 1977). The more rapid uptake of water by legumes is due to their larger embryo and the greater water-absorbing capacity of the embryo in comparison to that of the grass seed which is predominantly endosperm tissue (Fig. 2.1).

It is possible to stop imbibition of the seed and hold it in a 'primed state of suspended germination' (Heydecker, 1978). 'Primed' seed can be obtained by germination inhibitors, exposure to temperatures that are too low or too high for germination, hypertonic osmotic solutions (e.g. salt solution) or exposure to a wetting and drying cycle. The advantages of sowing primed seed lie in the reduction in the time between sowing and emergence (by as much as six days) and in the primed seed having more synchronous germination. The use of primed seed may, but does not always, increase seedling number and dry weight. Seed priming occurs in natural seed populations where exposure

Fig. 2.9. Cumulative germination percentages after 15 days from sowing for the rangeland species *Danthonia caespitosa* (Wallaby grass) and *Atriplex nummularia* (oldman saltbush) as a function of matric potential and temperature. (From Sharma, 1976.)

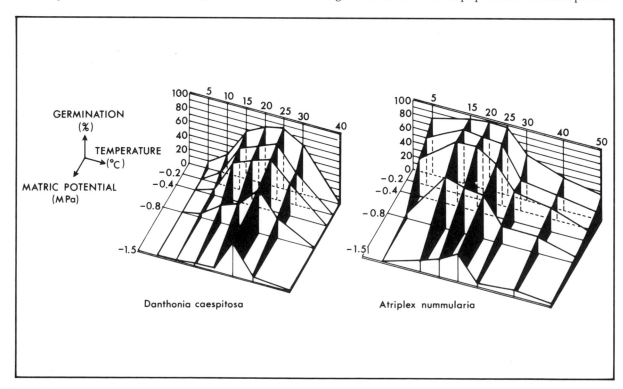

to wetting and drying may reduce the requirements for light and temperature for germination (McKeon, 1984).

The rate of germination and, for that matter, developmental events dealt with in Chapter 3 can be described by:

$$P = G\{1 - \exp[-k(t - t_0)]\} \qquad (2.3)$$

where P is the proportion of seeds germinated at time t, t_0 is the time until the first seed germinates, G is the maximum proportion to germinate (the germinability, Table 2.1) and k is a measure of synchrony: the spread in the population of time to germination. When k is high the slope of percentage germination versus time is steep, i.e. there is little spread in the population of time to germination (Milthorpe & Moorby, 1979).

In general, the effect of any suboptimal environment, e.g. temperature, on any event (germination, emergence, establishment and the components of vegetative generation) will cause G to be reduced and the median time for the event to occur ($1/t \times A/2$) to be delayed. Furthermore, the spread in the population about the median time will be increased, i.e. k will decrease.

In the case of temperature, it is generally recognized that there are high temperature (T_m) and low temperature (T_b) limits above or below which seeds do not germinate, and an optimum temperature for germination (T_o). The optimum temperature is usually not a single point: it is the temperature range where the germination rate is at a maximum. These temperature parameters vary not only with species, variety, genotype, origin and degree of maturity of the seed, but also with the physiological state of the seed prior to germination.

Germination characteristics such as the speed of germination and spread in the population respond curvilinearly to temperatures around the optimum (e.g. Hill, Pearson & Kirby, 1985). However, if there is a linear relationship between germination and temperature between T_b and T_m, as in tropical species (Garcia-Huidobro, Monteith & Squire, 1982*a*), this allows calculation of the speed of germination in terms of thermal time or accumulated temperature ('heat units'), according to:

$$1/t = \text{constant} \times (T - T_b) \qquad (2.4)$$

where T is the daily temperature and $T - T_b$ is the number of heat units (°C) accumulated during a day. For example, for pearl millet (*Pennisetum americanum*) for which T_b and T_m are 12 and 48°C respectively, the speed of germination was related directly to the number of heat units and it was found to be faster when day/night temperature amplitudes were 8°C

than under near-constant conditions (Garcia-Huidobro *et al.*, 1982*b*). Fluctuations in daily temperature improve germination (e.g. Lovato, 1981).

In addition to temperature, light, including daylength, water and oxygen are necessary for germination. The response to temperature may be altered by dormancy (Vegis, 1964), seed moisture content, light and treatment with nutrients and osmotica, e.g. nitrate. Interactions of this type are illustrated by Fig. 2.9 for the germination response of two semi-arid grassland species to the interaction between temperatures and matric potential which, with osmotic potential, largely determines soil water potential.

2.5
Vegetative generation
Two types of vegetative generation can be distinguished:
  (i) that where the next generation is initially attached to the parent plant; and
  (ii) that where the next generation is separate, i.e. dispersed from the parent plant.

These types are not mutually exclusive. They are common where the parent plant forms rhizomes or stolons. Rhizomes are underground stems; stolons are tillers in grasses or stems in dicotyledons which spread at an obtuse angle to the main stem they subtend, often running along the surface of the soil, and forming roots at their nodes.

Generation attached to the parent plant appears as a 'front' advancing from a centre of parent plants. When the stems and roots of the first generation die the second generation may re-invade this area or the hollow centre may, because of changed microclimate or fertility, become invaded by another species. For example, the stoloniferous legume white clover (*Trifolium repens*) is involved in a complex regeneration cycle. Initially, white clover and perennial ryegrass cohabit; as the nitrogen level rises due to dinitrogen fixation by the rhizobia attached to the roots of the white clover, ryegrass increases its dominance; it in turn is replaced by taller grasses such as *Dactylis glomerata* and as nitrogen status declines white clover again re-invades (Turkington & Harper, 1979; Fig. 3.12).

Vegetative generation which is initially attached to the parent plant is obviously effective only over short distances, causing the spread of a species rather than its initial colonization. It is most important in species which form prostrate stems which root quickly, e.g. white clover. White clover plants form many branches per unit area but many initially die (Chapman, 1983). A high population of apical meristems (4000 per m^2) maintains a high branch population despite the fact that few branches develop fully; this morphological characteristic improves this species' resistance to heavy grazing. In species such as white clover there are two generation pathways, namely seed and rooted stolons known as ramets. Seed of white clover is of importance in rapidly colonizing new habitats that exist after some form of soil or sward disturbance; vegetative generation is important in maintaining daughter plants of the same genotype in the changing micro-environment and species composition of long-term grasslands (Burdon, 1983).

Climatic factors and management influence generation pathways. Examples are (i) in wet years in subtropical Australia 54 per cent of white clover seedlings produced rooted stolons before death of the parent taproots compared with 6 per cent in dry years (Jones, 1980) and (ii) vegetative shoots and stolons of *Holcus lanatus* (Yorkshire Fog) increase under regular grazing or cutting (Watt & Haggar, 1980).

Vegetative generation which involves dispersal from the parent plant causes both short- and long-distance spread. At some distance from the parent plants, long-distance spread is called colonization. Vegetative dispersal requires that segments of the parent plant, usually a piece of stem containing one or two nodes, will form roots, commonly at nodes. The rate of dispersal depends on the ability of the parent plant to produce tillers or stems; the rate at which the stems break (preferably at the nodes when pulled by grazing herbivores) and detach from the parent; and on patterns of grazing. Presumably much of the spread of such segments is through grazing rather than by natural detachment and dispersal by wind or water. Two examples of vegetative dispersal are the grasses *Imperata cylindrica* and kikuyu (*Pennisetum clandestinum*). Imperata is widely distributed throughout Asia and Africa on land previously under crop or forest (Hubbard, 1944) and kikuyu, a native of east Africa between 1900 and 2700 m altitude, has been introduced to more than 20 countries in all continents (Mears, 1970). Imperata, although of some feed value when young, is considered a weed; kikuyu, although sometimes a weed, is more often sown (generally as vegetative pieces) as an improved pasture.

Sowing of species which depend on vegetative, not seed, generation, is labour-intensive or requires specialized machinery. Stems are cut or chopped so that pieces contain 1–3 nodes and these are planted into cultivated soil. For example, 450 kg of cut pangola grass (*Digitaria decumbens*) is sufficient to sow 1 ha of land. Planting is commonly done by broadcasting the pieces and then disc harrowing, or planting by hand into trenches. It is generally recommended that pieces are planted at a density of 1.2 per m^2 (e.g. Mears, 1970). Despite the relative slowness and often high cost labour component of vegetative generation some taxa, e.g. kikuyu and bermuda grass (*Cynodon* X) are propagated primarily or exclusively by this method. Vegetative propagation is also used for seed-producing species where quality seed is not available.

2.6
Further reading

Harper, J. L. (1977). *Population Biology of Plants.* 892 pp. London: Academic Press.

Roberts, H. A. (1981). Seed banks in soils. *Advances in Applied Biology,* **6**, 1–56.

Thompson, P. A. (1981). Ecological aspects of seed germination. *Advances in Research and Technology of Seeds,* **6**, 9–42.

3

Vegetative growth

Growth is the change, usually an increase, in biomass. We talk of the growth of a leaf, a plant or a grassland. Grassland growth is equivalent to productivity and is measured as tonnes of herbage per ha per year. Growth proceeds at a rate which varies over time and for a duration which depends on the life cycle of the plants and on their environment and management. Development, on the other hand, refers to the passage of an organism through its life cycle. This passage may be considered at the organelle or organ level or for the whole organism or plant community (Fig. 1.1). The life cycle of a herbage plant is divided into two developmental phases: vegetative and reproductive, according to whether meristems are seen to be producing further leaves or flowers respectively (Fig. 3.1). This superficial classification is widely used in grassland agronomy and it is often the basis of management, e.g. hay-making at 'the beginning of the reproductive phase' (one-tenth bloom in lucerne). However, Fig. 3.1 shows that there are a number of developmental events within the life of a herbage plant. This chapter discusses the development of vegetative organs and their rate of growth; Chapter 4 deals with the development of flowers and seeds.

Grasslands may be classified into five broad groups according to the developmental patterns of the main species in them:

(i) Annuals, in which the major component of the grassland is generated from seed and resown each year. This group includes sterile genera, e.g. some tetraploid ryegrasses (*Lolium* spp.) and species which are not permitted to set seed because they have a life cycle which is too long to allow them to disperse viable seed in a particular environment or because they are managed (e.g. grazed) so as to avoid seed production and seedling regeneration.

(ii) Self-regenerating annuals. This group includes species such as subterranean clover (*Trifolium subterraneum*) and many annual grasses in native grasslands which complete their life cycle and produce seed within one growing season and the dispersed seed germinates and re-establishes the grassland in a subsequent season.

(iii) Biennials. This is a loose term used to encompass species such as *Lolium multiflorum* in which most parent plants live for two seasons and maintenance of these species in the grassland requires management to ensure seed dispersal from the original plants or, more commonly, resowing. Lenient grazing may cause biennials to act as:

(iv) Short-lived perennials. Under these conditions the grassland is comprised mostly of species which regenerate vegetatively and to some extent from seed, but where the regeneration of the major species is not sufficiently robust to maintain the composition of the grassland beyond 3–5 years. Short-lived perennials, e.g. *Trifolium pratense* and in some areas lucerne, *Medicago sativa*, are therefore often used for hay, silage or intensive grazing until their degeneration necessitates resowing or, often, sowing of a grain crop.

(v) Perennials. Grasses such as perennial ryegrass (*Lolium perenne*) in temperate areas, paspalum (*Paspalum dilatatum*) or bahiagrass (*P. notatum*) in the subtropics and signal grass (*Brachiaria decumbens*) in the tropics, and legumes such as *Trifolium repens* and *Medicago sativa* may, depending on grazing management and fertility, commonly perennate for 5–20 years. Perennation is achieved by maintenance of the original rootstock, by natural vegetative propagation (Section 2.5) and to a relatively small extent by replacement of old plants by cohorts of seedlings.

Fig. 3.1. Developmental events (a) and growth (b) during the life of a herbage plant. Nitrogenase activity refers to upper nodules of an annual legume. The Development index is a scalar ranging from 0 at emergence to 1 at maturity. (Adapted from Summerfield & Wein, 1980, and Bergersen, 1982.)

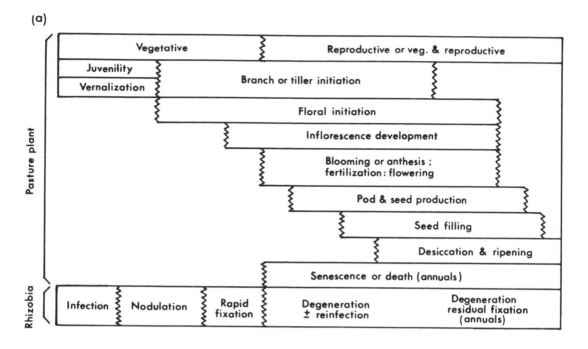

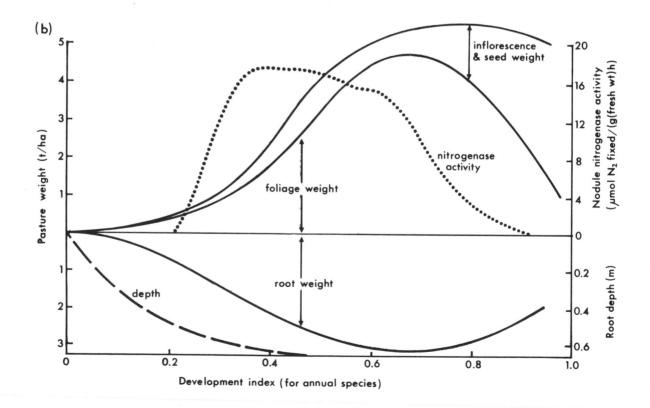

A species may be found in more than one grassland group because developmental patterns change with environment, e.g. lucerne, as shown above, and white clover (*T. repens*), which is usually perennial except in the subtropics, where it behaves as an annual.

3.1
Emergence and establishment

Germination finishes with the protrusion of the radicle from the seed coat (Section 2.4). Emergence is defined here as being completed with the penetration of the shoot above the surface of the soil; it is thus a property of the growing medium. The speed and effectiveness (or success) of emergence depend on seedling or emergence type, seed size and genotype, soil type, the environment and seed depth. The depth of sowing is considered in Section 2.2.

3.1.1
Seedling type

As early as 1885, botanists recognized that the shoot develops in one of two ways (Klebs, 1885, cited by Vogel, 1980):

  (i) epigeal emergence, where cotyledons are carried above the soil; or
 (ii) hypogeal emergence, where cotyledons stay in the soil within the seed coat (Fig. 3.2).

Fig. 3.2. Development of seedlings: epigeal (a) and hypogeal (b) emergence of a dicotyledon and hypogeal emergence of grass (c).

Within the epigeal group, which is exclusively dicotyledonous, Bekendam & Grob (1979) recognize two further categories: one where no epicotyl elongation occurs during early development, the other where the epicotyl is more or less elongated (Fig. 3.3). There are various categories of hypogeal emergence. The most common is the pattern typical of grasses in which the one cotyledon remains near the seed, below the soil surface, and the root system is comprised to a varying extent of seminal ('seed') roots (Fig. 3.3). Grass radicles enter the soil surface and penetrate within soil more easily than do legume radicles because the grasses have more root hairs, a smaller root diameter and a more acute angle of radicle entry into the soil (Campbell & Swain, 1973*a*).

3.1.2
Seed size and genotype

Large seeds give rise to more vigorous seedlings than do small seeds if they are of the same ploidy. Seeds of a higher level of ploidy, e.g. tetraploid seeds, are more vigorous than diploids of the same genus. Hill *et al.* (1985), studying eight temperate grasses at five temperatures found:

$$G = -3.5 + 0.81C + 0.36T - 0.0077T^2 \qquad (3.1)$$

where G is the growth rate (μg per seedling per day), C is a caryopsis factor to account for differences in vigour attributable to perenniality and ploidy (a score of 2 for an annual, 1 for a perennial, multiplied by 2 for

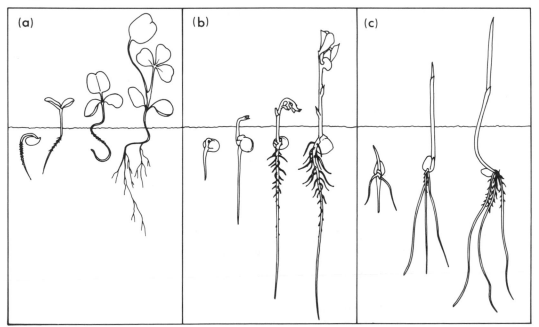

doubled chromosomes and 1 for a natural chromosome complement) and T is the mean daily temperature between 15 and 30°C. In another study which measured the area of cotyledons of herbage legumes during the twenty days following emergence, large-seeded legumes, e.g. lucerne and red clover, had more rapid emergence and faster growth than the small-seeded alsike clover and white clover although *Lespedeza*, which has intermediate seed size, showed the slowest growth (Hur & Nelson, 1985). Genotypic differences in the rate of cotyledon expansion, the ratio of embryo weight to endosperm or seed weight, the leaf area/weight distribution, the photosynthetic rate and differences in the onset and rate of leaf production all contributed to differences in early growth among species.

3.1.3
Seed bed

Densely-textured soils present a high resistance to the elongation of the seedling. Likewise, surface crusts and seals reduce the speed of emergence and decrease the percentage emergence of seedlings even when the environment is otherwise favourable. Arndt (1965) showed that, as the soil surface is wet, drains and then dries, its upward impedance (the resistance it would offer to a seedling attempting to break through the surface) increased two- to three-fold during wetting and up to nine-fold relative to dry soil as the soil was dried from field capacity to a 6 per cent

water content. Soil obstruction may be the greatest single cause of seedling mortality before establishment in rangeland situations. For example, Leslie (1965), with 17 sowings of tropical grasses, found the greatest cause of mortality (about 40 per cent) was soil impedance; this was aggravated by light penetration down cracks in the clay causing young seedlings of Rhodes grass (*Chloris gayana*) to develop a crown as much as 1 cm beneath the soil surface.

Deep planting and non-optimal temperature or soil water content will reduce the rate of hypocotyl elongation and delay emergence and thereby increase the likelihood that the seedling will be subjected to attack from pests or diseases.

3.1.4
Establishment

Plant establishment is here defined as the production of the first true leaf of the seedling. This, or any similarly arbitrary definition of establishment, purports to estimate the success of new seedlings in gaining 'ecological space': a share of the radiation, water and minerals which have to be divided among individuals within the whole sward. Thus plant establishment can be distinguished from the establishment of the grassland as a whole. Grassland establishment is an all-embracing term to cover agronomic practices related to sowing and seedling development from imbibition to a point where the seedlings have a high probability of survival, i.e. the sown seedling population has stabilized.

Establishment, as with emergence, depends on a complex of genetic, environmental and positional influences. In establishment it is important whether the emerged seedling is close to an already-established plant. If it is, then the growth of the new seedling will

Fig. 3.3. Categories of grassland species according to seedling morphology and early growth: the primary division is on the basis of cotyledon location determined by hypocotyl elongation (epigeal) or lack of elongation (hypogeal). (Adapted from Bekendam & Grob, 1979.)

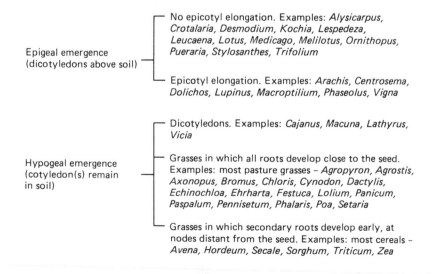

be retarded by shading, likely drying of the soil surface as the available water is used by the more prolific root system and possibly allelopathy. Allelopathy is a term used to describe deleterious effects that one plant can exert on its neighbour through complex carbohydrates being released adjacent to roots or from decomposing herbage residue. Dead plant material on the soil usually assists seed establishment (Evans & Young, 1970). However, nitrophilous weeds once killed collapse to form a mat on the surface of the soil and this reduces germination of herbage species, probably due to the weeds containing toxins which are leached into the soil (McCalla & Haskins, 1964).

As Harper (1977, p. 171) put it, the chance of a seedling surviving and being productive is greatest if it '(a) established before its neighbours, enabling it to pre-empt resources, (b) is well separated from its neighbours and (c) has weak neighbours'.

The speed with which seedlings emerge and become established can be described by equations such as Eqn (2.3). Predictions of cumulative percentage emergence with time are customarily used for even-aged populations, e.g. those seeds that are all dispersed or planted or watered, on one day. The effectiveness of a particular species in developing in a certain situation can be described mathematically. For example, Italian ryegrass (*Lolium multiflorum*) may have higher final emergence (*A* values) and faster emergence (a higher *k* value (Eqn 2.3)) than fescue (*Festuca arundinacea*) because of genetic factors and the ryegrass having a larger seed than the fescue (Hill *et al.*, 1985).

Each developmental event depends on the seedling having successfully passed the preceding event: establishment is only possible after the seedling has emerged, and emergence requires successful germination. In a grassland comprised of more than one species, the proportion of a certain species which completes a developmental event, say establishment, will depend on (i) the composition and the ratio of species at the end of the preceding developmental event, say emergence, and (ii) the effectiveness (or success) of one species relative to other species during current development. Mathematically this can be expressed as:

$$Y_m = \frac{a_m \times y_{(m-1)}}{y_{(m-1)} \times a_{(m-1)} + 1} \qquad (3.2)$$

where Y_m is the output, say the proportion of plants of one species at the end of a particular phase of development (say, vegetative competition) in year m expressed in terms of the output of the previous phase, $y_{(m-1)}$ (say, establishment). The *a* values are expressions of the effectiveness of one species relative to that of

others during, say, vegetative competition. Torssell, Rose & Cunningham (1975) used this analysis to show that in a grassland comprised of tropical grasses and legumes, the ratio of legume plants established relative to grass plants varied from less than 1:1 to 167:1 over five years.

Consequently, species composition is often highly variable at the end of establishment of sown or self-regenerating species, particularly where other plants are present in the grassland. Species composition depends on the number of seedlings established in any year. However, this number is not explained simply by any single factor such as the size or number of seeds (e.g. Smith, Biddiscombe & Stern, 1972). The number of germinable seeds depends on the dynamics of the seed bank (Section 2.3). Studies in mediterranean environments show seed production from self-regenerating annuals may be halved if the length of the rainy season is reduced by three weeks (Collins, 1981) and that an increase in the stocking rate reduces seed production and increases dispersal (Gramshaw & Stern, 1977*a*). The number of emerged seeds depends on a complex interaction involving seed placement (which is rarely measured), seed size, hardseededness, genotype and the micro-environment. Grazing, trampling and uprooting of plants are further factors which affect establishment, although their importance is difficult to assess in rangelands (Chapter 7).

3.2
Forms of development

Fig. 3.2 shows the early development of dicotyledons which exhibit epigeal and hypogeal emergence. In these plants, which include all important herbage legumes, the apex of the stem is raised as internodes elongate. Thus the apex occupies an exposed position above the last lateral branch. By contrast, in grasses the apex is in a protected position, encircled by leaf sheaths and near the soil surface until internode elongation occurs, commonly at the time the apex becomes reproductive.

The site of potential growing points is significant for adaptation to grazing, fire and climatic extremes. Examples of adaptation are found in the tropical legumes *Centrosema virginianum* and *Stylosanthes guianensis* var. *intermedia*. Cold tolerance by both is related to their ability to regenerate from growing points below the soil surface. In *Centrosema* growth occurs from the cotyledonary node which in accessions from colder climates remains below ground and thereby protected from frost (Fig. 3.4); there is a negative correlation between its survival in the cool tropics and the height of its cotyledonary node (between 0 and 2 cm above ground) because temperatures below −2°C at the node kill the plant (Clements & Ludlow, 1977). In *Stylosanthes*, regrowth comes from the protected hypocotyl. Regrowth after fire and grazing are dealt with in Sections 3.4.2 and 3.4.3 respectively.

3.2.1
Leaves

Leaf development is reviewed by Dale & Milthorpe (1983b).

Rates of leaf initiation and appearance increase with increasing irradiance (Dale & Milthorpe, 1983a). Leaf expansion ceases first at the leaf tip and margins and last at the leaf base, in both grasses and dicotyledons. The rate of expansion of individual leaves is highest 2–5 days after emergence and declines thereafter. During the period of fastest expansion, rates are highly sensitive to temperature, ranging in ryegrass (*Lolium temulentum*) from 0.08 to 1.8 mm per h at 2–20°C (Thomas & Stoddart, 1984). Rates of leaf expansion are highest in spring when plants are reproductive (Chapter 4).

3.2.2
Branches

Branching or tillering takes place when meristems in the axils of leaves undergo rapid cell division and cells behind the meristem elongate to raise the branch beyond the surface of the parent stem. In dicotyledons the branches may be prostrate or horizontal, as in most temperate clovers, or erect, formed at an acute angle to the main stem, as in some tropical legumes and

Fig. 3.4. Position of growing points affects tolerance to grazing, fire and climatic extremes. *Centrosema virginianum* lines which originate in the warm tropics have a growing point (cotyledonary nodes) above ground whereas lines from cool climates have nodes below ground and greater tolerance to frost. (From CSIRO, 1979.)

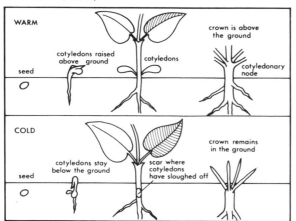

shrubs. In grasses, the tiller meristem is initially encircled by the sheath of the leaf. The elongating tiller then emerges from the sheath in one of two ways. Most commonly in turf and tussock grasses, the tiller elongates between the sheath and the true stem or sheath of the next-formed leaf until it appears near the base of the lamina of the parent leaf. Less commonly in grasses, the tiller penetrates the parent sheath and gives rise to a rhizome or stolon. A rhizome is an underground horizontal stem which bears scale leaves and buds and sometimes roots, at nodes. A stolon or runner is a horizontal, above-ground stem which forms roots. Subterranean clover (*Trifolium subterraneum*) has above-ground, prostrate stems which do not form roots (Fig. 3.5b) whereas white clover

Fig. 3.5. Plant morphology. (a) White clover, showing its stolons; (b) subterranean clover with horizontal stems from which peduncles elongate to push the fruit into the soil; (c) perennial ryegrass with upright tillers; (d) kikuyu, which has rhizomes.

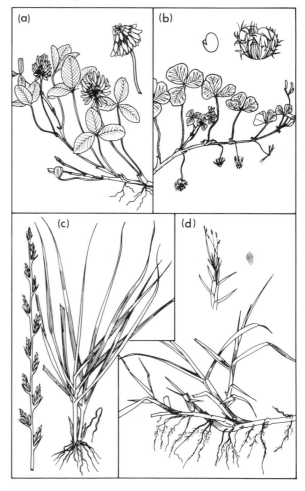

(*Trifolium repens*) has stolons (Fig. 3.5a). Erect temperate grasses such as ryegrass (*Lolium* spp.) show the common pattern of the grass tiller elongating with its leaf sheath (Fig. 3.5c). Prostrate grasses such as kikuyu (*Pennisetum clandestinum*) are prostrate because the tiller penetrates the leaf sheath and forms rhizomes (Fig. 3.5d). Between these extremes are erect species such as fescue (*Festuca arundinacea*) and paspalum (*Paspalum dilatatum*) which have short rhizomes beneath the soil.

Branching or tillering shows distinct seasonal patterns. In England, tillering of temperate grasses is highest in summer (Fig. 3.6), when temperatures are close to the optimum for tiller production and when soil water and radiation are plentiful. The temperature optimum for tillering is the same or as much as 5–10°C below the optimum for growth. Branching or tillering is sensitive to radiation and to light quality: it is increased by exposure to red light (600–700 nm wavelength) at the end of the day (e.g. Deregibus, Sanchez & Casal, 1983). The onset of tillering is delayed by shading and low nutrition (Gregory & Sen, 1937; Fletcher & Dale, 1974).

3.2.3
Roots

Roots form directly from the embryo (the seminal root system) and from nodes of stems and stolons. The root systems of herbage plants were described in classical works by Weaver (1926) and Troughton (1957). Root elongation during the seedling stage is usually faster than stem elongation. There is generally a functional relationship between root mass and shoot mass (Brouwer, 1962). This allows empirical descriptions of root mass in relation to shoots, for particular species and locations. For example, in England Burns (1980) proposed that the effective rooting depth (D, cm) could be related to shoot weight (W_s, t per ha) and plant population (p, plants per ha):

$$D = 6.1W_s + 1.5 \times 10^{-5}p + (1.8 \times 10^{-3}/a^2) - 2.1 \tag{3.3}$$

where a is the average radius of the roots (cm).

The extent and shape of the root system vary according to soil characteristics such as texture, bulk density (and void ratio), aeration, soil water potential and fertility. Taylor & Gardner (1963) considered the interaction between root elongation and the physical properties of the soil and showed that root penetration decreased with increasing bulk density (reduced void ratio) and additionally that penetration rates at any bulk density were higher in wet than in dry soil. Some generalizations (A. P. Hamblin, 1985, personal communication) are:

(i) Roots cannot grow into rigid pores narrower than their own diameters.
(ii) Root tips can exert pressures up to 10 MPa to expand non-rigid pores in deformable (plastic or friable) soils.

Fig. 3.6. Tillers in a timothy sward in England. The dashed line shows the change in the total number of tillers; vertical distances between continuous lines represent the number of tillers produced in the period between samples (dates given). The fate of tillers present on a particular sampling date can be gauged by following the continuous lines. Note that the rate of tillering is highest in summer. (From Jewiss, 1972.)

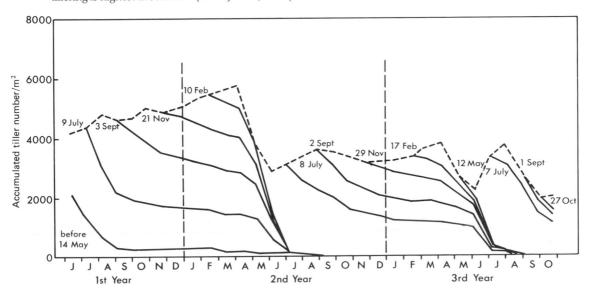

(iii) Critical values at which root growth ceases in compact soils vary from less than 1 to greater than 4 MPa depending on the soil composition, plant species and pore water potential.

(iv) Roots growing through apedal sands must expand pores or be deflected, and roots growing in coarse-structured fine-textured clays are inevitably restricted to regions of lowest resistance, i.e. they grow in major crack planes.

(v) Roots grow downwards (geotropically) and root elongation rates are reduced substantially (to a half or less) if they are constrained to grow down channels at more than 45° to the vertical.

(vi) High levels of mechanical impedance cause root tips to become buckled, with an expansion of the cross-sectional diameter and a proliferation of root hairs behind the meristematic zone; in these circumstances the respiration rate increases but cell expansion is reduced.

Generally, root growth will be severely impeded at bulk densities of 1.55, 1.65, 1.8 and 1.85 g per cm^3 in clay loams, silt loams, fine sandy loams and fine sands respectively (Bowen, 1981; Jones, 1983). Therefore root length density, the length of roots per unit volume of soil, varies enormously between species and location. Total lengths of up to 17 km per m^2 of ground area or densities of 10 cm per cm^3 in topsoil have been recorded for annual legumes (Pearson & Jacobs, 1985). In perennial grasslands, root densities in topsoil in the United Kingdom are 35–100 cm per cm^3 (Garwood & Sinclair, 1979); Milthorpe & Moorby (1979) present a root profile for tall fescue in which root densities declined from 63 cm per cm^3 in the topsoil to 3.6 cm per cm^3 at 50–60 cm depth.

Legume roots may be infected by bacteria of the genus *Rhizobium* to form nodules. Nodules vary in appearance and size. The bacteria within these nodules fix nitrogen from the soil atmosphere and the amino acids and amides which are produced are translocated in the phloem of the host plant, largely to the plant shoot. The significance of rhizobia in nitrogen nutrition of grasslands is considered in Section 5.2. Nodule weight reaches a peak or plateau one-half to two-thirds of the way through the life of an annual legume; the rate of nitrogen fixation by the nodules may decline appreciably before foliage and seed growth stop (Fig. 3.1). The development, form and effectiveness of nodules is reviewed by Bergersen (1982).

Roots may also be infected by fungi which grow on the root exodermis and between and within cells of the root cortex. Those which penetrate within root cells (the endotrophic mycorrhizae) are significant for mineral, particularly phosphorus, nutrition of the grassland (Section 5.2). Other organisms, e.g. *Azospirillium*, live freely on or within the root. Free-living organisms have no direct effect on structural development although they do affect growth by making variable quantities of minerals highly available for uptake by the plant.

3.3
Growth

Growth depends on development although the relationship between them is not simple. Growth of both individual plants and the grassland generally follows a sigmoid curve (Fig. 3.1) from the time of establishment until annuals die or a 'steady state' is reached in perennials when the canopy is intercepting all incoming radiation and the growth of new leaves, stems and roots is approximately equalled by the death of old organs. Growth after grazing (regrowth) is often linear (Section 3.4). The rate of growth depends on:

(i) the interception of radiation;

(ii) genetic differences in the utilization of radiation, which may be related to photosynthetic pathway and plant and sward structure, i.e. development;

(iii) carbon balance: the balance between the net carbon dioxide exchange rate and daily respiration and death;

(iv) environment: the extent to which water, temperature, soil structure and fertility are suboptimal; and

(v) management, including grazing management (Section 7.1) and the incidence of pests and diseases.

3.3.1
Interception of radiation

The radiation which is available for grassland growth depends firstly on the amount of solar radiation at the top of the atmosphere. This changes in a seasonal pattern according to latitude. It is attenuated by depth of atmosphere, clouds and particulate matter, leading to a certain amount of net radiation on the herbage. Of this, 50 per cent is photosynthetically active radiation (I), having wavelengths between 400 and 700 nm. Formulae for calculating average net photosynthetically active radiation, hereafter abbreviated 'radiation', according to latitude and day of the year, are given in Appendix A.

The growth rate of grassland is described by εI where ε is the efficiency of conversion of solar radiation to carbohydrates and I is the radiation. This assumes that the herbage is intercepting all the radiation and that other factors, e.g. temperature, are optimal. This allows us to predict that potential productivity is highest at low latitude (Section 1.2) although seasonal cloud cover reduces this potential. Where less radiation is intercepted, growth will be reduced. According to Warren-Wilson (1971):

$$G = \varepsilon I[1 - \exp(-KL)] \qquad (3.4)$$

where G is the growth rate (kg per ha per day), I is the radiation (MJ per ha), ε is the photosynthetic efficiency (kg per MJ) and the expression $1 - \exp(-KL)$ is an approximation to the fraction of radiation intercepted by a canopy with leaf area index L and an extinction coefficient K. The leaf area index is the ratio of leaf area (one side only) to ground area; K is the ratio of leaf area when it is projected onto a horizontal surface (the ground) to the total leaf surface area.

This correlation between growth rate and the fraction of radiation intercepted by the grassland holds for long periods of time, such as weeks between grazings, so long as the type of radiation, specifically the proportions of sunny and cloudy weather, remains constant. Over short periods of time, canopy photosynthetic rate is proportional to the fraction of intercepted radiation when either (i) all leaves are working on the linear part of the photosynthesis–radiation response curve, such as happens in the early morning or late afternoon or on a cloudy day, or (ii) photosynthesis of shaded leaves which are situated low in the canopy is negligible compared with the photosynthesis of sunlit leaves, as occurs on days of high radiation (Monteith, 1972).

Fig. 3.7. Growth and canopy architecture. (a) Growth after emergence or defoliation is initially slow if the canopy extinction coefficient (K) is low due to relatively low light interception, whereas after canopy closure the growth rate is greatest if the canopy is erect, i.e. where K is low. (b) Extinction coefficients and light interception (shown here on a logarithmic scale) differ between closely-related selections of ryegrass (*Lolium perenne*) S231 due to (c) differences in canopy structure. Ryegrass type 5 had long leaves and erect tillers whereas ryegrass type 10 had short lax leaves and prostrate tillers. (From: (a) Monteith, 1981; (b) and (c) Rhodes, 1971.)

There is an asymptotic relationship between the fraction of radiation which is intercepted and the product of K and L. This draws attention to the interrelationship between the extinction coefficient and the leaf area index. Erect canopies with vertical leaves, e.g. grasses, in which K may be about 0.3, require a substantially higher leaf area index for complete interception of radiation ('canopy closure') than do prostrate canopies, e.g. legumes, where K may be 0.7–0.9. Grasses may intercept virtually all (95 per cent) radiation at leaf area indices of 6–9 whereas temperate legumes will do so at leaf area indices of 2.5–4.

During early growth, such as when grasslands are regenerating from a small leaf area or from seed, a low extinction coefficient is disadvantageous, as shown in Fig. 3.7a. This represents foregone production. During grassland establishment, production can be increased only by high seeding rates, altering planting patterns, e.g. row spacings, sowing species in which the seedlings have a high relative growth rate (which is usually correlated with large seed size) and minimizing competition between the sown or desired regeneration species and other species by planting in a favourable climatic and cultural situation.

The advantage of vertical structure after canopy closure is that radiation is attenuated more gradually as it moves down into the pasture canopy compared with rapid attenuation in a prostrate (high K) canopy. Thus, more leaf area can be 'held' on the canopy above the point where daily leaf photosynthesis equals respiration. This point is noticeable: below it, leaves do not contribute to net growth, they appear chlorotic (yellow to white) and leaf senescence is accelerated.

The 'holding' of a large leaf mass has managerial advantages, principally, that the interval between grazing can be relatively long and the amount of feed which can be removed at a single grazing is large. The canopy with largest leaf area at complete radiation interception may also have the highest growth rate after canopy closure. One English study found that under infrequent cutting the most productive *Lolium*

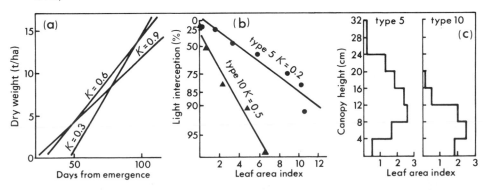

genotypes were those that had low extinction coefficients (high leaf area indices at complete light interception); however, under frequent cutting, highest productivity came from genotypes with high extinction coefficients and a large proportion of leaf at the base of the canopy (Fig. 3.7b, c). Elsewhere, in fescue (*Festuca arundinacea*) there was no relationship between growth rate and extinction coefficient of the sward although the extinction coefficients ranged from 0.3 to 0.6 (Sugiyama *et al.*, 1985).

Differences in the extinction coefficients of dissimilar species such as grasses and legumes are widely recognized. Such differences also exist between closely-related genotypes, e.g. within *Lolium* (Fig. 3.7). These may be related in part to small differences in development pattern (as leaves age they become more planophile), to genetic differences in leaf angle, and to the structure of the plant, e.g. stemminess.

3.3.2
Utilization of radiation

The potential photochemical efficiency of a leaf is about 12 g carbon dioxide per MJ of radiation (e.g. Charles-Edwards, 1982). This is usually taken as being equivalent to 8 g dry matter (DM) per MJ since 44 g carbon dioxide are converted through photosynthesis into 30 g of carbohydrate. The energy costs of synthesis vary according to the compounds which are produced. Herbage plants, including noxious species, contain a bewildering variety of compounds (Chapter 6). The major compounds are carbohydrates, proteins and lipids. These have energy values of about 17, 24 and 37 kJ per g respectively. Thus, if herbage were to contain carbohydrates only, then the photochemical efficiency might be 8 g per MJ but if they contain 12 per cent protein and 2 per cent lipids then it would be 7.4 g per MJ and for legumes containing 35 per cent protein the efficiency would be 6.6 g per MJ.

Actual photochemical efficiencies usually range from 1.5–2.5 g per MJ to as high as 5 g per MJ. This range appears to be due largely to:

(i) Photosynthetic biochemistry. Tropical grasses have a combination of the Hatch-Slack (C_4) photosynthetic pathway, located in the bundle sheaths which surround the phloem, and the Calvin (C_3) cycle, which is located in the leaf mesophyll cells. This combination appears, at least under high radiation or temperatures, to be intrinsically more efficient than the C_3-cycle alone, which occurs in all legumes and temperate grasses.

(ii) Regulatory factors within the plant which are related to the stage of development and which cannot be wholly explained by changes in sward structure. These factors may simply be changes in the partitioning of carbohydrate to allow development of greater leaf area expansion in preference to leaf thickening or growth of other organs, or a result of changes in respiration. Whatever the cause, the apparent photochemical efficiency, as estimated by growth rate, under what appears to be an optimal environment, may increase by 40 per cent (Anslow & Green, 1967) to as much as 400 per cent (Fig. 3.8a) during inflorescence elonga-

Fig. 3.8. Seasonal growth rates are related to environment, management and plant development. (a) Ryegrass grassland, showing higher growth rates in spring than at other times of the year. Values are mean growth rates of shoots on two farms in Bay of Plenty district, New Zealand. Dec indicates December: months of sampling follow a clockwise sequence. (b) Winter growth depends on the time of first floral primordium production in a number of temperate grasses at Orange, Australia. ((a) Adapted from Field, 1980, by Kemp, 1984; (b) from Kemp, 1984.)

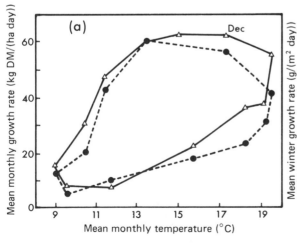

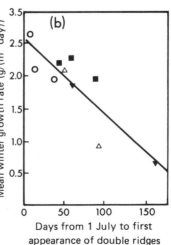

tion and emergence. Thus, it is possible to get a correlation between growth rate and the time when the plant apex changes from a vegetative to a reproductive form (Fig. 3.8b).

(iii) Herbage structure, particularly as this affects light interception (Section 3.3.1). At establishment, differences in growth rate among species are related to rates of leaf expansion. However, after the beginning of tillering, high relative growth rates are correlated with high rates of tillering (Sambo, 1983). The structural causes which correlate most highly with growth differ between species: path coefficient analysis indicates that tiller production might contribute most to rapid growth in Italian ryegrass (*Lolium multiflorum*) whereas leaf expansion is more important in determining the rate of growth of *Bromus catharticus* and fescue (*Festuca arundinacea*) (Hill *et al.*, 1985).

Values of 2–5 g dry matter per MJ of intercepted radiation are equivalent to growth rates of 250–625 kg per ha per day where incoming radiation is 350 MJ per m^2 per day, a level which would saturate photosynthesis for a 10 h day. For comparison, shoot growth rates of temperate pastures, e.g. *Lolium*, rarely exceed 200 kg per ha per day in Europe and New Zealand whereas maximum short-term growth rates of tropical grasses, e.g. *Pennisetum americanum* may be as high as 580 kg per ha per day (Pearson, 1984).

3.3.3
Carbon balance

The carbon balance is the balance between photosynthesis and respiration which leads usually to a daily net gain in dry matter, i.e. growth (e.g. Pearson & Hunt, 1972b; Pearson, 1979). The growth rate of a sward or whole plant is not related simply to its photosynthetic rate (e.g. Rhodes, 1972; Nelson, Asay & Horst, 1975). This is because at the sward level, growth is determined more by length of time when conditions are favourable for development than by photosynthesis (Monteith & Elston, 1983) and when considering the individual plant it is respiratory losses, the pattern of utilization of carbohydrate and senescence which have more effect on net carbon gain than does photosynthesis.

Respiration causes a 20 to 40–60 per cent loss of daily photosynthate. The lower value is obtained from measurements of net gas exchange by single plants and microswards in controlled environments. The loss of potential efficiency is usually about 50 per cent (Robson, 1982) but may occasionally be as high as 80 per cent in the field, where it is difficult to apportion losses between respiration and death and detachment. More mechanistically, respiration can be related to previous production of photosynthate (the so-called 'synthesis component' of respiration) and to plant age, leaf area and protein content (the 'maintenance component'). That is:

$$r = a \times P_N + b \times W \tag{3.5}$$

where r is the rate of respiration, P_N is the amount of prior daily photosynthesis, W is the dry weight of grassland and a and b are constants. The coefficient of synthetic respiration is 0.25–0.3 and it is independent of temperature. The coefficient of maintenance respiration increases with temperature, most probably in exponential fashion because its rate doubles with every 10°C increase in temperature. In ryegrass in England, synthetic respiration varied from 0.38 of total dark respiration in autumn to 0.46 of total respiration in spring (Parsons & Robson, 1982).

Legumes may have different respiratory loadings compared with grasses on account of the carbon used in supplying energy for dinitrogen fixation. The extent of the 'cost' of dinitrogen fixation is contentious. It is considered further in Section 5.2; here it is sufficient to note that nodules may respire about 10 per cent of the gross photosynthate (Minchin & Pate, 1973).

Leaf death reduces productivity. Leaf death also affects herbage quality and animal intake (Section 6.4). The beginning of leaf death occurs earlier at high than at low temperatures: in cowpeas (*Vigna unguiculata*) the beginning of death was inversely correlated with mean air temperature (a correlation r^2 value of 0.9 (Littleton *et al.*, 1979)) although in subterranean clover leaves started to die at 28 days after sowing irrespective of temperature, plant size and structure (Fukai & Silsbury, 1976). The rate of death and detachment is considered to increase linearly with increasing temperature: rates in the field commonly range from 4 per cent per month in winter to 30 per cent per month in summer.

3.3.4
Water availability

Water, temperature, soil structure and fertility affect directly the rate of grassland growth. They also affect growth indirectly through their influences on development. A detailed discussion of these environmental factors is beyond the scope of this book because each herbage type has a unique sensitivity to environment and there is interaction between the environmental factors. Here we draw attention to some principles.

Water is particularly important in determining the growth of grasslands in climates with distinct dry seasons such as the savannah grasslands of the wet-and-dry tropics. Thus Coe, Cumming & Phillipson

(1976) were able to relate grassland growth and the weight of large African herbivores (W, kg per km^2) to average annual rainfall (R, mm per year):

$$W = 8.68R - 120 \qquad r^2 = 0.88 \qquad (3.6)$$

(where r^2 is the percentage of variance accounted for by the regression) in twelve ecosystems receiving less than 700 mm of rain.

Analysis of the amount of soil water which is present and available to the grassland requires a consideration of the amount of rainfall, the water-holding capacity of the soil and the losses of water from the herbage and soil. Available soil water is the volume of water held between field capacity and wilting point, i.e. with an energy of between −0.03 and −1.5 MPa, in the root zone. This varies five-fold according to soil texture (Fig. 3.9). Root penetration and the total volume of available soil water for the plants are greatest in soils which have a medium texture, low bulk density and good aeration.

Grassland growth is reduced where available soil water falls below 25 per cent of the maximum or when the evaporation from the herbage cannot be met by movement of water to the roots and through the xylem to the leaves. Transient water deficits occur in plants where evaporative demand is high, usually about solar noon in high radiation environments. Such transient deficits usually cause stomatal closure and are seen as leaf flaccidity, folding or drooping. They are more common and severe for species with shallow roots where capillary rise of water within the soil cannot keep pace with the depletion of water in a relatively small shallow soil zone. Thus midday wilting due to loss of turgor, particularly of petioles, may occur in clover but not be apparent in grasses within legume/grass mixtures.

Species which have the same relationship between growth and percentage available soil water will nonetheless show different sensitivities to low water supply because of variation in rooting depths: the total volume of exploitable soil water is a characteristic of species, age and environment. For example, increasing either temperature or stocking rate will increase the partitioning of photosynthate to shoots relative to roots; this may make herbage more liable to water stress.

Furthermore, some grassland species may evaporate more water per unit of dry matter accumulated than others. Thus, although several species may have the same relative sensitivity to available water, the species which uses more water per unit dry matter increment has a lower Water Use Efficiency (WUE) and may more quickly deplete soil water and therefore show diminished growth. Differences in WUE on a leaf or plant basis indicate that species with the C_4 photosynthetic pathway are more efficient than those having C_3-photosynthesis and of temperate origin. However, apparent WUE on an annual basis shows a ten-fold range from 0.002 to 0.02 g dry matter per g of water, irrespective of photosynthetic biochemistry, in semi-arid grasslands (Table 3.1). It is in these grasslands where we might expect differences in WUE to confer ecological advantage to the more efficient species.

If water deficits are imposed gradually, plants may adapt metabolically by synthesizing osmotic sol-

Fig. 3.9. Available soil water. (a) The amount of water held between wilting point (WP) and field capacity (FC) varies between soils: horizontal bars show the total amount held in four different soils. (b) Relative root growth as influenced by water potential, and the interaction between water potential and air-filled porosity and soil strength (resistance to penetrometer) for a podzolic soil. SAT, saturation. (From: (a) Norman et al., 1984; (b) Cornish, 1986.)

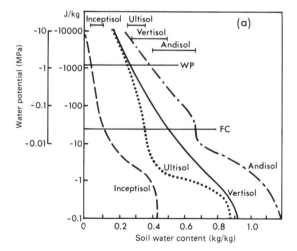

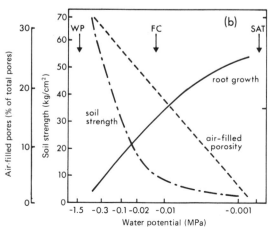

utes within their leaves. This osmotic adjustment increases with time under stress and, together with a decrease in the elasticity of cell walls, may confer some tolerance to declining soil water. However, the current view is that osmotic adjustment in semi-arid grassland species is of small significance: it provides 'adjusted' species with only one or two days' extra growth before wilting occurs (Wilson & Ludlow, 1983).

Non-lethal water stress also causes morphological changes such as leaf shedding and reduction in leaf size. These reduce the rate of water loss (and the rate of growth) of the 'adjusted' herbage. This in turn provides extra time for growth before soil water is reduced to the point at which wilting occurs.

3.3.5
Temperature
Temperature affects the growth rate of grasslands by affecting separately each process of development and the rates and directions of the metabolic pathways associated with growth. Species may be grouped into four broad types according to the responsiveness of their growth rate to temperature (Fig. 3.10). This is a gross, but agronomically useful, simplification. There are also important differences in temperature responsiveness within, as well as among, genera. Effects of

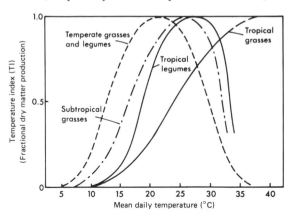

Fig. 3.10. Temperature index (TI): herbage growth expressed as a fraction of the growth potential of the species at optimum temperature. Grassland species are grouped into temperate grasses and legumes, tropical legumes, subtropical grasses and tropical grasses. (Adapted in part from Fitzpatrick & Nix, 1970.)

Table 3.1. *Net primary productivity and water use efficiency in ungrazed temperate and tropical grasslands*

Location	R (mm) (A)	Annual green shoot biomass production (t/(ha year)) (B)	Annual root biomass production (t/ha year)) (C)	Apparent WUE $(B + C)/A$ (t DM/t water)
Temperate grasslands				
Colorado, USA	232	10.1	42.2	0.0225
Bouteloua gracilis (C_4)				
Buchloe dactyloides (C_4)				
Texas, USA	355	17.6	43.6	0.0172
Bouteloua gracilis (C_4)				
Aristida longiseta (C_4)				
South Dakota, USA	360	18.8	24.8	0.0121
Agropyron smithii (C_3)				
Stipa viridula (C_3)				
Oklahoma, USA	674	28.6	25.6	0.0080
Andropogon geradi (C_3)				
Panicum virgatum (C_3)				
Tropical grasslands				
Queensland, Australia	664	12.2	6.0	0.0027
Thyridoleips mitchelliana (C_3)				
Monachather paradoxa (C_3)				
Queensland, Australia	647	15.4	16.5	0.0049
Cenchrus ciliaris (C_4)				

Source: Various authors, in Christie (1984).

temperature on growth and development have been reviewed by Field, Pearson & Hunt (1976) for lucerne and by McWilliam (1978) and Humphreys (1981).

3.3.6
Nutrition

The effects of nutrition on growth are considered in some detail in Chapter 5. Ion uptake depends in part on the concentration of elements at the root surface (Section 5.2) so where elements are not readily available in the soil, growth will be retarded. Mengel & Kirkby (1979) describe soil and plant analyses which assess the availability of various elements. The extent to which growth is affected by the concentration of a mineral is given by an asymptotic relation:

$$G = 1 - b \exp(-CX) \qquad (3.7)$$

where G is the growth rate scaled between zero and 1 (maximum growth for a particular grass) C is the coefficient of curvature of the response curve, b is an empirical constant used to describe the responsiveness of the site, ranging from 0 to 1, and X is the concentration of the element, which may of course be varied by applying fertilizer. The species which is best able to grow at low mineral concentrations will have a competitive advantage over other species within the grassland; this leads to over-shading and further unequal allocation of water and minerals, and to changes in the composition of the grassland (Section 3.6).

3.4
Regrowth

The utilization of plant growth involves some form of defoliation. Consequently vegetative growth depends heavily on the mechanism of regrowth and the factors which affect it, namely fire, grazing, invertebrates and diseases.

3.4.1
Mechanism

When herbage plants are grazed or cut, their rate and form of regrowth depend on:
 (i) whether or not the apical meristem is removed;
 (ii) the levels of carbohydrates within remaining organs;
 (iii) the potential rates of photosynthesis of leaves and stems which were previously shaded;
 (iv) root mass and activity; and
 (v) the environment, particularly temperature.

If the apical meristem remains intact, leaf production continues, but if the meristem is removed then lateral tillering or branching is necessary before the resumption of leaf production. Thus, the height of grazing or cutting changes the location of shoots which support regrowth. Cutting near to ground level produces buds on the stem base and rhizomes; cutting higher leads to axillary buds on stems which in turn suppress the outgrowth of buds beneath them (Table 3.2).

In a study of the physiology of regrowth of *Pennisetum*, a tropical forage grass, tillers with apices regrew faster and elongated more than decapitated tillers where the apex was removed (Muldoon & Pearson, 1979). The faster regrowth of intact tillers was related to an initial decrease in stem weight and loss of a relatively high proportion of soluble carbohydrate out of the stem; translocation of more carbohydrate from the stem to new growth; and some mobilization of less soluble carbon (proteins and polysaccharides) from the old stem and leaf sheaths. Decapitated tillers regrew mostly by producing branches in axils of leaves, as found by Belyuchenko (1977) (Table 3.2).

The concentration of carbohydrate in the organs remaining after defoliation may be important in determining the rate of regrowth because respiration, especially of roots, increases following defoliation and

Table 3.2. *Height of cutting of* Pennisetum purpureum *determines the percentage composition of bud types on regrowth*

Type of bud	Height of cutting (cm)					
	0	5	10	15	30	50
Rhizomatous	95	90	29	3	4	0
Crown basal shoot	5	6	25	31	10	2
Apical bud below cutting height	0	0	9	12	18	20
Axillary bud on vertical shoot	0	4	36	55	69	78

Source: Belyuchenko (1977).

small amounts of carbon move into the tissues which grow in the first 5–10 days after defoliation (e.g. May, 1960; Ehara, Yamada & Maneo, 1966). Roots do not generally provide substrate for new growth (Marshall & Sagar, 1965).

Increased respiration and lack of photosynthate appears to be the reason for a decrease in root weight (volume and length) and cessation of root extension, for up to 20 weeks after defoliation, so that the distribution of roots with depth changes following grazing (e.g. Pook & Costin, 1971). In legumes, there is loss of nodules and a fall in nitrogen fixation (Ryle *et al.*, 1985).

Net photosynthetic rates of old, previously-shaded leaves increase when these leaves are exposed to full sunlight. The photosynthetic rate of the canopy shortly after defoliation depends on the age of these lower leaves: the rate declines as the leaves age and beyond a certain age these leaves are unable to re-attain their young, potential rate when they are exposed to full radiation. Hodgkinson (1974) showed that the decline in photosynthetic rate with age is related to leaf position, the amount of radiation it receives while it ages and the presence of new growth elsewhere on the plant.

In the 1960s there were some studies which suggested that regrowth would be fastest if defoliation was lenient and swards were maintained at leaf area indices of about four. Grazing experiments do not universally support this. Increasing the stocking rate is as equally likely to cause increases as decreases in net productivity (Section 7.2).

King *et al.* (1984) found that the rate of regrowth of ryegrasses (*Lolium perenne*) which are adapted to various defoliation treatments increases with increasing leaf area index, i.e. regrowth is faster under lax grazing. However, the tiller density decreases and senescence increases with increasing leaf area index. Thus King *et al.* (1984) found that net productivity was constant when leaf area indices ranged from 2 to 4.5 and harvestable herbage was greatest when leaf area indices were between 2 and 3. In other studies, ryegrass subjected to hard grazing at a stocking rate of about 48 sheep per ha to maintain the leaf area index at 1 had a canopy photosynthetic rate which was only two-thirds of that of canopies which were leniently grazed at a leaf area index of 3. Under hard grazing a higher percentage of new growth was harvested by the sheep so that there were fewer losses due to death and decay in the sward which had a leaf area index of 1 than in the sward held at a leaf index of 3 (Table 3.3).

In summary, it seems that residual leaf area is the factor which is most important in determining regrowth but that we still lack a mechanistic understanding which would allow us to devise optimum management strategies for grazing under seasonally-changing conditions (Simpson & Culvenor, 1986).

3.4.2
Fire

Burning of old grass has been accepted for two hundred years as a way of improving herbage quality and sometimes increasing productivity (Thunberg, 1793). Fire is responsible for maintaining the botanical composition of the great grasslands of the world, such as the savannahs of East Africa and northern Australia; protection from fire and grazing leads to changes in botanical composition, particularly the invasion of herbaceous and woody species and reduction in grass cover. Regular but not necessarily annual burning of grasslands probably improves productivity: removal or killing of trees leads to increased productivity from native grasslands (Gillard, 1979) although slight increases in productivity may lead to only marginal, or no, increases in stocking rates in extensive grasslands. West (1965) reviews the role of fire in the management of grasslands. Fire causes:

(i) The removal of unpalatable shoots carried over from previous growing seasons.

(ii) The stimulation of new growth by increasing the amount of radiation reaching the bases of plants

Table 3.3. *Regrowth of ryegrass pastures in England as affected by grazing. Values are means of 14-day periods over two years*

Grazing	Sheep/ha	Daily photosynthesis (kg OM[a]/(ha day))	Respiratory loss (% of photosynthesis)	Grazing intake	
				kg OM[a]/(ha day)	% of photosynthesis
Lenient	21	300	31	38	13
Hard	48	209	34	53	25
Overgrazed	125	166	34	99	63

[a] OM, organic matter.
Source: Leafe & Parsons (1983).

and thereby causing tillering, the rate of which is accelerated by the heat and the release of minerals in the surface soil.

(iii) The stimulation of flowering and seed production in some species.

(iv) The control of invasion of herbaceous and woody plants. Woody species such as shrubs on which animals browse are susceptible to burning and grazing at the seedling stage; however, some herbaceous species have protected buds (e.g. *Centrosema*, Fig. 3.4) and eucalypts, which have ligno-tubers, are highly adapted to fire;

(v) The destruction of parasites, e.g. ticks, which transmit diseases of livestock.

Prescribed burning is inexpensive and it may be effective in controlling weeds or woody species, but it may result in little or only a seasonal increase in grassland growth (Hamilton & Scrifes, 1982) as estimated by a 17 per cent return to investment in burning in a Utah study (Ralphs & Busby, 1979).

3.4.3
Grazing

Grazing management affects the form of regrowth which in turn affects photosynthetic rates and carbon balance (Table 3.3). These, together with selectivity of grazing and uneven return of nutrients to the grassland influence the competition between species and the composition of the grassland (Section 3.6). The effects of grazing strategies are considered in Chapter 7.

3.4.4
Invertebrates and diseases

Elton (1930) drew attention to the fact that the biomass of invertebrates exceeds the weight of domestic animals grazing on a grassland. He used the term 'ecological pyramid' to describe the organization of grassland (at the base), invertebrates, vertebrate herbivores and carnivores (at the apex of the pyramid of biomass). There are numerous insects which live in the soil, on roots, on the soil surface or on foliage. Macfadyen (1971) gives examples of their abundance: e.g. in one square metre of temperate grassland there might be 10^{12} bacteria, 10^6–10^9 nematodes and 10^3 of each of slugs, worms, mites, Collembola (spingtails), beetles, millipedes and fly larvae. They may eat live herbage, dead standing herbage, or graze off organic matter and other organisms. Their role in litter decomposition is reviewed by Dickenson & Pugh (1974), and their significance in mineral cycling within grasslands is discussed in Chapter 5.

Insect populations reflect the availability of feed and suitable microclimate. Generally, temperate grasslands are more likely to be damaged severely by insects than are pastures in tropical or arid regions, because in temperate regions there are more improved grasslands, more diverse fauna and climates which permit the maintenance throughout the year of large populations of insects (Wallace, 1970). Thus, grassland management, in that it affects the food supply, also affects insect populations. For example, pasture cockchafers may be the red-headed genera (*Adoryphorus couloni*), which feed on roots, or the black-headed foliage feeders (*Aphodius* spp). In southeast Australia, cockchafers are native, but they have emerged as serious pests only when pasture improvement, using superphosphate fertilizer and sowing subterranean clover, has increased the availability of live feed and litter. Litter is necessary for the survival of large numbers of first instar larvae (Maelzer, 1962), as well as being an alternative source of food for the adult cockchafer.

The amount of grass which is eaten by pests depends on the population density, feeding habit, age and diet selection of the pest and on the temperature at feeding, as discussed by Rodell (1978) for a temperate North American grassland. Pests may be allocated to two broad classes according to their feeding habit:

(i) Those that graze herbage or litter and remove a constant amount without affecting the ungrazed material. Such insects, grubs, etc., might, for example, graze lateral roots or leaf tips. They will destroy a grassland only if the density of pests is so high that their consumption exceeds the herbage growth rate.

(ii) Pests which, by grazing, have a denuding or deleterious effect on the remaining plant, e.g. by severing tillers or leaves at ground level or removing plant apices. Here, the amount of herbage killed is not related simply to the intake of the insect or grub.

Such categories are helpful although they are not strictly adhered to. As Table 3.3 shows, management affects the photosynthetic rate and subsequent growth rate and the amount of available feed; these in turn result in substantial variations in the percentage of productivity which is 'left over' after livestock grazing for invertebrates. This left-over feed and the attendant variation in microclimate in turn affect the diversity of invertebrates, their populations and their feeding habits. The porina grub (*Wiseana* spp.), which is prevalent in New Zealand, illustrates this change in feeding habit according to feed availability. In grassland which was closely grazed to maintain a sward height of 1 cm, the apparent removal of herbage was 5.9 mg per porina per day and, due to a denuding grazing habit (type (ii) above) the apparent removal per porina increased with increasing herbage growth rate and availability up to about 2 t per ha. Above 2 t per ha, however, the herbage loss per porina decreased to 2 mg per day and the damage they caused, e.g. the percentage loss of herbage cover, also decreased (Barlow, 1985). The authors of this research concluded that the porina pest is a denuding type which would cause greater revenue losses at high stocking rates than at low ones.

The effects of pest density on grassland growth may be generalized (Fig. 3.11). This includes the possibility of growth enhancement, compensation, a proportional decline in growth with increasing pest population, and decreasing losses per pest due to competition between pests at high pest populations. Not all parts of Fig. 3.11 will apply to every grassland system because of the diversity of pests and variations in timing of pest attack relative to plant development. Enhancement is probably uncommon and the lower limit to herbage growth may be zero where the pests completely defoliate or kill the pastures.

In addition to affecting directly the growth of a pasture, pests may reduce productivity by changing the botanical composition. Changes in botanical composition may in turn lead to losses of soil fertility, decreased animal production and soil erosion. Pests cause changes in botanical composition by affecting the competition between plant species through selective removal of leaf area, growing points and roots, by selectively affecting seed production or by denudation until one species is killed in preference to others. Plant pathogens, e.g. the fungus *Collectotrichum gloeosporioides*, which causes anthracnose, have the same effects: selective reduction in seed yields are shown in Table 3.4.

The impact of pests and diseases is mediated by site-specific factors such as environment and soil fertility. Turf-grass diseases, e.g. those caused by *Corticum fulciforme*, are diminished at high levels of potassium fertilizer and at high levels of nitrogen, but not in the presence of phosphorus (Goss, 1972). Such complex interactions make it difficult to generalize about the likely number of invertebrates within grasslands or to predict the thresholds at which chemical control of the invertebrates becomes economic. These interactions do, however, leave open the possibility of pest management using biological methods. Roberts (1979), trying to control scarab larvae, hypothesized that alternating each year from relatively low to high stocking rates was unlikely to allow environmental conditions to remain favourable for a sufficient time for any of the scarab species to increase to damaging numbers. Four years of Roberts' experiments showed that alternating grazing pressure can reduce the density of scarab larvae, but this method still needs to be tested commercially. Likewise, severe grazing of dryland lucerne can give high levels of control, in excess of 95 per cent reduction in numbers of legume aphids (Allen, 1984). Other cultural methods for the control of insect pests include the addition of fertilizer to increase the fitness of grassland plants so that they can compensate for pest damage; strategic ploughing or cultivation to remove the food of pests, to crush insects and pests or to expose them to desiccation and natural enemies (e.g. Birks & Allen, 1969); and manipulation of crop rotations to discourage pests which are monophagous, i.e. dependent on one source of feed (e.g. Sandow, 1983).

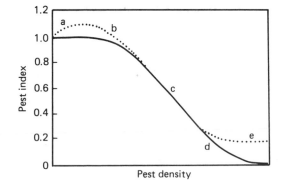

Fig. 3.11. Pest index (PI): effect of pest incidence on grassland growth. A speculative graph to show (a) phases of growth enhancement, (b) compensation, (c) linear decline, (d) pest competition and (e) low (tolerance) level. (Adapted from Fick, 1984.)

Table 3.4. *Effect of anthracnose on dry matter yield and seed yield of* Stylosanthes *spp., showing that selective damage varies between pasture species. All values are t per ha*

	Sprayed (no anthracnose)	Unsprayed (natural infection)
Dry matter yields[a]		
Townsville stylo	3.90	1.84
Fitzroy	10.52	2.27
Verano	10.79	7.20
Graham	13.91	7.44
Seed yields[a]		
Townsville stylo	1.23	0.19
Fitzroy	1.15	0.18
Verano	1.25	0.61
Graham	1.37	0.59

[a] Dry matter and seed yields were obtained from different locations.

Source: R. D. Davis, Personal communication (1985).

3.5
Competition

The extent to which one species intercepts more than its fair share of resources, and possibly kills its neighbour, has been termed its 'competitiveness'. In large part competitiveness can be explained by differences between individual plants of one species and between genotypes in the utilization of resources: different responses to temperature, more upright habit leading to over-shading and proportionately greater interception of light so long as grazing pressure is low, deeper rootedness, etc. Allelopathy also contributes to interactions among plants although its quantitative importance seems to vary greatly between species.

3.5.1
Thinning

When considering populations which become established at about the same time, i.e. age cohorts, such as occurs when a grassland is sown or when there is a definite start to the growing season, there is a negative linear relationship between mean plant weight and plant density (Harper, 1977). That is, there is a tendency for some plants to die as their neighbours increase in weight. Density-related mortality may not occur until some weeks after establishment; its onset is earlier at high population density and it may start earlier in some species than in others. Density-related mortality will occur earlier in *Trifolium pratense* than in *Medicago sativa* at the same population density (Malberg & Smith, 1982).

Density-related mortality occurs in mixed grasslands and in monocultures, where it is called self-thinning. Where it occurs regularly, as in evenly-spaced seedlings, it is described by:

$$w = a \times p^{-1.5} \tag{3.8}$$

so that

$$W = w \times p = c \times p^{-0.5} \tag{3.9}$$

where w is the mean weight of an individual plant, W is the weight of the herbage, a and c are constants and p is the plant population (Yoda *et al.*, 1963).

Self-thinning has been reviewed by White (1982). It is unclear why it operates at such a constant rate and why for some species self-thinning within a species is more important than interference between species. Harper & McNaughton (1962) found that in mixed populations each species reacted more strongly to its own density than to the density of its companion species. By contrast in *Rumex*, a weed, thinning may respond more to alien than to like plants (Cavers, 1963). Thinning is usually explained by shading between neighbours whereby the weaker neighbour receives progressively less light. It is not clear to what extent thinning is due to depletion of soil water and nutrients.

3.5.2
Competitiveness and growth rate

There are many studies of interactions between plants during vegetative growth using replacement series. In these experiments, two or three species are planted in mixtures of varying proportions and measurements are made of the yield of the individual components and of the total mixture as the proportions are changed. These studies are reviewed by Trenbath (1978). For example, soil nitrogen may frequently be a limiting resource in semi-arid grassland. It is therefore instructive to study the interactions in artificial mixtures of co-occurring C_3- and C_4-grasses at various levels of nitrogen. Christie & Detling (1982) examined the effects of nitrogen and temperature on mixtures of *Bouteloua curtipendula*, a C_4-prairie grass, and *Agropyron smithii*, a C_3-species. At a day-night temperature regime of 20/12°C the C_3-species became dominant when there was direct competition for soil nitrogen, whereas at a temperature regime of 30/15°C the situation was reversed. Grasslands comprised of sown legume, sown grass (*Phalaris aquatica*) and unsown species, mostly native and weedy taxa, on soil of low native fertility become dominated by the weeds if there is no fertilizer or only nitrogen fertilizer is added, whereas the grasslands tend to grass dominance under high fertilization (Wolfe & Lazenby, 1973). Many more examples are given in Harper (1977) and in Wit, Tow & Ennik (1966).

Short-term studies, particularly with seedling plants in replacement series, equate competitiveness and success with high rates of growth. In some situations, however, high growth rates will be disadvantageous. Summer dormancy, as found in ecotypes of grasses of Mediterranean origin but not in cool temperate ecotypes, is more desirable than summer growth for grasses which need to survive extreme heat and intermittent drought; these grasses show a negative correlation between survival and growth over the summer (Silsbury, 1961). High rates of growth may also be negatively correlated with quality (Chapter 6). There is a tendency for decreased growth at cool temperatures when subtropical grasses acclimate to locations of higher latitude (Pearson *et al.*, 1985). Likewise, the frost resistance of temperate grasses is negatively correlated with the ability of the grass to grow at low temperature (Cooper, 1964).

The discussion above has emphasized grasslands in which seedlings make a major contribution to the total plant population and thus to composition and productivity. By contrast, grasslands comprised of perennial species will have a seedling component but, depending on species, this may be of much less significance than the perennating plants. In a grassland dominated by perennating plants such as *Phalaris aquatica* or *Lolium perenne* and *Trifolium repens* in cool temperate climates, the seasonal changes in the percentage composition of available dry matter depend on the seasonality of growth of the component species. This gives rise to growth being comprised of different proportions of grass and legume in each season. Legumes generally have prostrate, planophile canopies and shallow roots, so they will contribute most to the total growth of the grassland under conditions of high available water or frequent irrigation and high grazing pressure. Among temperate grasses, tall fescue (*Festuca arundinacea*) and phalaris (*Phalaris aquatica*) dominate over perennial ryegrass (*Lolium perenne*) during periods of water shortage due to the smaller root system of *Lolium*.

The greatest complementarity between species (mixtures out-yielding the best species in monoculture) appears to occur among species having different thermal responses. Thus in subtropical climates irrigated mixtures of C_3- and C_4-species are dominated by C_4-grasses in summer and by C_3-grasses and legumes in winter. There are very few examples of complementarity where the components have the same thermal responses and canopy architecture. In one situation where grass mixtures did out-yield the best monoculture after three years of grazing, the out-yielding was attributed to complementary root growth: a triploid *Lolium* hybrid took up more nutrients from deeper soil when grown with *Lolium* or *Festuca* than when grown alone (Whittington & O'Brien, 1968).

3.5.3
Weeds
The effects of weeds on the growth of herbage plants has not been comprehensively researched or reviewed. Competition between weed and herbage species follows the principles outlined as operating among herbage species, i.e. unequal sharing of limited resources which leads to one component directly and deleteriously affecting other components. Weeds may directly reduce herbage growth through over-shading and competition for water and nutrients; they make herbage less accessible to livestock and they reduce livestock production through their deleterious effects on the digestibility of the feed eaten by the grazing animals.

Because the direct effect of weeds on herbage growth may be less significant than their effect on accessibility, removal or control of weeds may have varying results. Removal of prickly pear (*Opuntia polyacantha*) may not increase the growth of its co-habitant blue grama (*Bouteloua gracilis*) but it makes the grass more accessible to cattle (Bement, 1968). Leafy spurge (*Euphorbia esula*), which infests over 320 000 ha of North Dakota can be controlled with herbicides but Lym & Massersmith (1985) found that forage production, dominated by *Poa*, increased for only 27 of 59 herbicide treatments. In the other situations, eradication of weeds causes substantial increases in herbage growth. For example, eradication of velvet mesquite (*Prosopis jubiflora*) caused a three-fold increase in perennial grass production in Arizona (Cable & Tschirley, 1961).

Woody perennial shrubs invade under-grazed grasslands (e.g. Vorster, 1982; Roux & Vorster, 1983; Hennessy *et al.*, 1983). Shrubs such as mesquite (*Prosopis* spp.) and creosote bush (*Larrea* spp.) in the southwest USA, *Acacia* in Australia and karoo bush (*Eriocephalus* spp. and *Pentzia* spp.) in southern Africa probably directly reduce grassland productivity over large areas. It is now possible to kill these weeds using pellets (e.g. Herbel, Morton & Gibbens, 1985). It is not clear to what extent the advantages of shrub removal (in terms of increased herbage growth and accessibility) will be offset by loss of shade and soil stability.

Herbicides are used increasingly to manage the botanical composition of the grassland: an issue which is different from their use to control weeds. In one of many examples of manipulation of competition, Haggar & Bastian (1980) found grass-specific herbicides applied in late winter in England increased the proportion of clover and suppressed grasses. To increase clover above 20 per cent necessitated a 70 per cent reduction of grass growth in spring. However, this reduction was offset by later growth so that, although botanical composition was altered, total annual yields were largely not affected.

3.5.4
Long-term changes in composition
Long-term trends in grassland composition occur when factors affecting growth and regrowth (Sections 3.3, 3.4) consistently favour one species over others, or when one species will be favoured or seriously disadvantaged at a critical stage of its life cycle. For example, heavy grazing during flowering and before seeds are physiologically mature disadvantages one species relative to those which are not producing seed at the time of grazing (Section 7.1).

Changes in composition may sometimes be related to fertilizing. For example, fertilizing with nitrogen and phosphorus leads to dominance of

Bromus in Californian rangeland. The changes in composition may appear to be gradual and long term but they will often, even within long-term trends, show large changes in botanical composition over relatively short periods of time, e.g. within five years. Such long- and short-term changes were shown in a temperate grassland in Belgium in which *Agrostis tenuis* came to account for 80 per cent of the herbage after 14 years with no fertilizer whereas no species accounted for more than 30 per cent of the herbage when it was fertilized with phosphorus, potassium and calcium (Hecke, Impens & Behaeghe, 1981). The return of dung and urine also markedly affect grassland composition (Section 7.1).

A general model for long-term changes in species composition of a temperate grassland is shown in Fig. 3.12. The species of most importance will change between regions; those in Fig. 3.12 refer to Wales. *Agrostis tenuis* may invade swards in New Zealand (Harris, 1974) and *Bothriochloa macra* swards in southeast Australia (Cook, Blair & Lazenby, 1978). Nonetheless, two features are widely applicable: regeneration depends on soil factors as well as management and there is no stable climax association in the cycle.

Fig. 3.12. Postulated regeneration cycle within a permanent grassland in Wales. (From Turkington & Harper, 1979.)

3.6
Efficiency of net primary productivity

Net primary productivity (NPP) is the net production, or growth, of grassland per unit of time. It is expressed as kg or tonne per ha per year (Table 1.2). The efficiency of NPP is a measure of the efficiency with which the energy of incident solar radiation is converted to plant dry matter. The theoretical maximum efficiency of the NPP over short periods (days or weeks) is 12 per cent. Short-term values are usually in the range of 2 to 6 per cent during periods of active growth, as in spring. For example, lucerne swards intercepting 60–70 per cent of the radiation had efficiencies of utilization of intercepted radiation of about 10 per cent (6.2 and 5.1 g CO_2 per MJ radiation when watered weekly or at two-week intervals respectively). However, the efficiency of conversion of intercepted radiation into top growth was only 3 per cent or 2.1 and 1.5 g per MJ under the two watering regimes (Whitfield *et al.*, 1986).

Annual efficiencies of grassland growth are commonly in the range 0.1–2 per cent, largely because there are periods when the environment is far from optimal for growth and also when dormancy, death or defoliation cause the grassland to intercept little or no radiation.

The efficiency of NPP is usually higher in sown, highly managed pastures, where species have been selected for long duration and high rate of growth, than in natural grasslands. For example, in semi-arid

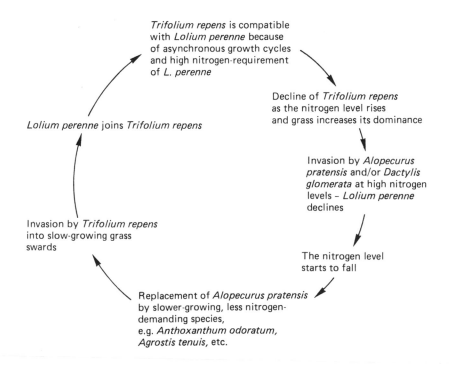

Trifolium repens is compatible with *Lolium perenne* because of asynchronous growth cycles and high nitrogen-requirement of *L. perenne*

Decline of *Trifolium repens* as the nitrogen level rises and grass increases its dominance

Lolium perenne joins *Trifolium repens*

Invasion by *Alopecurus pratensis* and/or *Dactylis glomerata* at high nitrogen levels – *Lolium perenne* declines

Invasion by *Trifolium repens* into slow-growing grass swards

The nitrogen level starts to fall

Replacement of *Alopecurus pratensis* by slower-growing, less nitrogen-demanding species, e.g. *Anthoxanthum odoratum*, *Agrostis tenuis*, etc.

southern Australia, a legume pasture stocked at 9 sheep per ha had an annual efficiency of NPP of 0.42 per cent in contrast to a native grassland with 2 sheep per ha for which the efficiency of NPP was only 0.14 per cent (Williams, 1964).

3.7
Further reading

Dale, J. E. & Milthorpe, F. L. (eds.) (1983*b*). *The Growth and Functioning of Leaves.* 540 pp. Cambridge University Press. Five chapters on growth; chapter by Monteith & Elston on foliage productivity.

Johnson, C. B. (ed.) (1981). *Physiological Processes Limiting Plant Productivity.* 395 pp. London: Butterworth. Chapters by Woolhouse, Monteith and Stanhill, although illustrated with crops, provide a theoretical background to grassland growth.

Pearson, C. J. (ed.) (1984). *Control of Crop Productivity.* 315 pp. Sydney: Academic Press. Three chapters on growth of grasslands and six on physiological processes which regulate productivity.

Robson, M. J. (1980). A physiologist's approach to raising the potential yield of the grass crop through breeding. In *Opportunities for Increasing Crop Yields*, ed. R. G. Hurd, P. V. Briscoe & C. Dennis, pp. 33–49, Boston: Pitman Press.

Wilson, J. R. (ed.) (1978). *Plant Relations in Pastures.* Melbourne: Commonwealth Scientific and Industrial Research Organization. Ecophysiology and analysis of interrelations between grassland plants; also six chapters on ecology of specific grasslands e.g. white clover-based pastures.

4

Flowering and seed production

Stability through persistence is a desired feature of natural grasslands and sown pasture. This is particularly true in grasslands comprised of one or more species which are intended to remain productive for several years. Reproductive development and the subsequent building up of seeds in the soil are essential for the persistence of grasslands based on annuals and they are an important means of providing new seedlings in some perennial grasslands (Fig. 2.2). An understanding of the factors which control flowering and subsequent seed production is desirable for (i) delineating environments in which sown selections are capable of producing seed for regeneration; (ii) adopting agronomic and grazing strategies to ensure successful flowering, seed production and sustained animal productivity; and (iii) selecting sites for commercial production of seed.

The phases encompassed in reproductive development frequently overlap (Fig. 3.1); their duration varies depending on their sensitivity to environment. If we understand the phases involved and their flexibility, we may be better able to manage flowering and seed production. This may be done by adjusting the rate and duration of developmental phases so long as these changes are free of compensatory changes in the rate or duration of other phases.

Seed is the product of interest: the first component of seed yield potential is the density of bud sites (plants per unit area × tillers or branches per plant, Chapter 3). Bud density is related to the length of the vegetative phase and thus to the timing of floral initiation (Fig. 3.1). The anticipated outcome of reproduction, the production of viable seed, is often expressed as the reproductive capacity (*RC*) of individual plants where:

$$RC = t \times f_t \times sp \times fl \times f_{fl} \times V_s \qquad (4.1)$$

t is the number of tillers per plant, f_t is the percentage of fertile tillers per plant, *sp* is the number of spikelets per ear, *fl* is the number of flowers per spikelet, f_{fl} is the percentage of fertile flowers (seed bearing) per spikelet and V_s is the percentage of viable seeds for, in this model, a grass plant.

This chapter considers juvenility before looking at flowering, seed formation, seed production and some of the implications of flowering for growth and management. It does not consider in detail the needs of commercial seed production.

4.1
Juvenility
Juvenility has commonly been defined as the early phase of growth from seed during which flowering cannot be induced by any treatment (Fig. 3.1). This definition does not include the possibility that some perennials may have a juvenile or pseudo-juvenile phase each year in individual shoots. For herbage grasses it is uncertain, although likely, that juvenility is a property of each tiller rather than of the whole plant (Felippe, 1979). Likewise there is circumstantial evidence that each individual branch of the perennial legume *Stylosanthes guianensis* (stylo) may have to pass through a juvenile phase in order for that branch to flower (Ison & Humphreys, 1984a). The juvenile phase is commonly short or virtually non-existent in ephemerals and many annual herbage plants but it may be very long in browse shrubs.

In most grasses juvenility may last from 10 to about 40 days but it may be as long as several years (Tainton, 1969). In legumes, juvenility also varies within and between species (Aitken, 1985; Ison & Hopkinson, 1985).

The mechanism of juvenility is not clearly understood. Current hypotheses suggest that the passage of juvenility is linked to changes in the hormonal balance due to extension growth or to the accumulation of a substrate (promoter and/or inhibitor) that is associated with growth. The presence or absence of juvenility, and its possible duration, are genetically controlled. The absence of juvenility is linked closely with 'early maturity' in plants. 'Early' cultivars generally produce more seed but less total dry matter than 'late' cultivars. Thus in autonomous (day-neutral, DN) flowering plants or where conditions are always inductive, a longer juvenile phase allows greater vegetative growth, sites for inflorescence initiation and prospects of increased seed yield. In contrast a long juvenile phase may make time of sowing critical in a species in which flowering is attuned to the regular, seasonal progression of daylength or temperature.

4.2
Morphological changes at flowering

Meristems are the sites of cell division and organ development. The terminal or apical meristem of the central axis, and subsequently of branches or tillers, differentiates an axillary meristem with each new leaf primordium. The number and location of meristems are critical in providing resistance to, and recovery from, grazing as well as in determining the potential sites for development of tillers and inflorescence primordia. The grass inflorescence is terminal on the shoot (a rare exception is kikuyu, *Pennisetum clandestinum*) whereas the first-formed inflorescence in legumes may be terminal or axillary depending on the species. Thus production of further leaf initials stops at floral initiation in grass tillers (Fig. 4.1a, b) and at most sites in determinate legumes; in contrast leaf production may continue after floral initiation in indeterminate species or in some determinate species due to continued sympodial branching from axillary meristems. In perennials a reservoir of axillary meristems or vegetative tillers provides the plant with the potential for continued vegetative growth following flowering.

Leaves are the sites of induction: perception of photoperiod (daylength) which leads to the production of unknown translocatable substance(s) capable of 'evoking' appropriate meristems (Evans, 1969). Leaves vary in their ability to perceive photoperiod. This depends on their age, position on a stem and area. The morphology of leaves changes frequently during induction. Evocation results in cellular changes at the apex which become evident as morphological changes (Fig. 4.1c); these changes are common to groups of plants (Moncur, 1981). At floral initiation apices have also increased in size (Fig. 4.1c). Floral initiation is usually definable by the formation of

'double ridges' or 'bulges' on the apex of grasses or legumes (Fig. 4.1c). Thus the inflorescence (head, spike, panicle) is initiated first, and then its component parts, which are of variable age and which have potentially different rates of development. Stem elongation, including the elongation of basal internodes of grasses, frequently accompanies floral initiation (Fig. 4.1b). The consequences of this are dealt with in Section 4.5.1. The patterns of morphological change are better documented for the winter cereals (Fig. 4.1) but these are likely to be comparable with those of many of the temperate pasture grasses.

Some meristems escape induction. The total number of floral primordia is thus determined early in reproductive development by (i) the number of axillary buds at the start of induction; (ii) the stimulated rate of primordia initiation after induction; and (iii) the rate of differentiation of induced primordia. These processes are interrelated: it has been suggested that a longer

Fig. 4.1. The pattern of morphological changes on the main stem during the change from vegetative to reproductive development of an annual grass (*Triticum aestivum*) sown in the autumn (northern hemisphere). (a) Number of ●, primary tiller buds; ○, primary tillers. (b) ●, Mean leaf number per plant; ○, stem length. (c) ●, Cumulative number of primordia; ○, number of florets on ninth spikelet. The diagrams summarize changes in the shoot apex appearance from the vegetative to spikelet primordia initiation stage. E, emergence; V, T, D, S, apices in a vegetative, transitional, double ridge and spikelet initiation stage of development respectively; A, anthesis. (Adapted from Baker & Gallagher, 1983.)

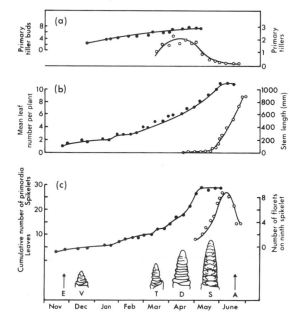

duration of induction may lead to an increase in the rate of primordia initiation and a decrease in the rate of floral differentiation, which may in turn lead to the production of more primordia (Ison & Humphreys, 1984*c*).

4.3
Flowering

Floral initiation may occur autonomously (day-neutral plants which flower without an external stimulus), as is the case with many arid zone or rangeland grasses (Fig. 4.2), or in response to environmental stimuli. Possible stimuli include daylength (photoperiod), temperature and water or nutrition stress; in some situations floral initiation may be controlled by interactions of photoperiod and temperature. Floral initiation may be associated with new source–sink relationships resulting in changes in photosynthate distribution, increases in growth rate, reductions in leaf-to-stem ratios; reduced tillering or branching (Section 3.2), decreases in the density of forage and thus in animal intake (Section 6.3), increases in concentrations of structural carbohydrate, which reduce quality (Section 6.1), and reduced nitrogen fixation. We do not have a clear understanding of the physiological processes leading to floral initiation; it is not certain if increased apex size *per se* leads to flowering, but critical physical size of the apical dome plus action of an endogenous inhibitor to primordium initiation has been used in a simple model of flowering (Charles-Edwards *et al.*, 1979). Several other models have been proposed.

4.3.1
Environmental controls of flowering

The time of flowering can be measured by the rate of development of the inflorescence (d*F*/d*t*) and it is related to the weather experienced by the grassland:

$$dF/dt = f(D, P_v, T, I) \tag{4.2}$$

where D is the daylength, P_v is the vernalization period, T is the temperature and I is the light intensity. Landsberg (1977) has used this relation as a basis for a model of flowering in *Lolium perenne*. Here it is a relevant framework within which to consider environmental controls of flowering in addition to a brief consideration of control by moisture and nutrition stress.

Response to photoperiod may be qualitative (obligate) or quantitative (facultative) (Fig. 4.2); most temperate pasture plants are long-day plants (LDPs) requiring long days (short nights) for floral initiation. This is consistent with their flowering in late winter-spring-early summer (Fig. 4.3). Many, but not all,

Fig. 4.3. Development pattern in three perennial and two annual grassland species at Melbourne, Australia (38°S). Sowing time was mid-summer (December) or later (shown by a vertical bar); later developmental events are flower initiation (■), anthesis or first flower appearance (●) and first ripe seed (▲). Contrasting cultivars of the annual legume *Trifolium subterraneum* are included. The lower figure shows photoperiod (○--○), and mean monthly temperature (●--●). (From Aitken, 1985.)

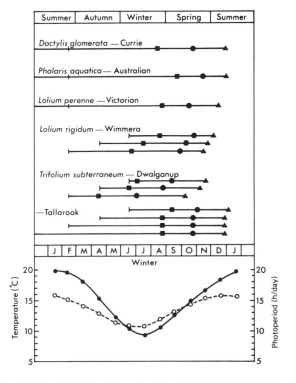

Fig. 4.2. Photoperiod response curves: effect of daylength on time to floral initiation for a quantitative SDP (▲), a qualitative SDP (●), a quantitative LDP (△), a qualitative LDP (○) and a day-neutral (DN) plant (□).

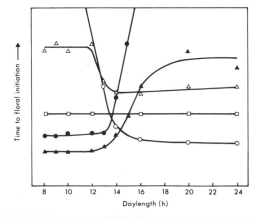

tropical and subtropical pasture plants are short-day plants (SDPs) (Ison & Hopkinson, 1985). Both SDPs and LDPs may flower at a common daylength (Fig. 4.2), the point being that LDPs will cease to flower, or flowering will be delayed, if daylength is further reduced, whereas SDPs show the opposite behaviour. More specialized photoperiod requirements have been identified: some ecotypes of white clover (*Trifolium repens*) are short-long-day plants, having an obligate requirement for short days before the long-day requirement can be met. In this type, floral initiation in the field in New Zealand occurs in November–December but stops in January (Thomas, 1980). The autumn-flowering *Stylosanthes guianensis* cv. Cook is a long-short-day plant with reverse requirements to those of white clover (Ison & Humphreys, 1984*a*). There is evidence, in need of clarification, to show that some plants respond to increasing or decreasing daylength rather than to duration of the day (or night) *per se*.

Plants with an obligate requirement for certain photoperiods will not initiate flowers until the critical photoperiod is reached, i.e. when the daylength is greater than the critical photoperiod for LDPs or shorter than the critical photoperiod for SDPs. Such plants will flower at the same time despite different sowing dates (Fig. 4.3). Variation in the critical photoperiod is the basis for differences in maturity among ecotypes of many species, e.g. subterranean clover (Fig. 4.3) (*T. subterraneum*) and *S. humilis*. Knowledge of this may be used to select herbage cultivars whose flowering season matches the expected length of the growing season (Section 4.5.3).

Many temperate herbage species are not 'competent' to respond to appropriate daylengths without prior exposure to a period of low temperature, i.e. vernalization. This is a term which now refers commonly to effects of low temperature (usually but not invariably in the range 0–10°C) which lead to both 'competence' to flower and to initiation.

There is considerable variation within and between species in their response to vernalization (e.g. Flood & Halloran, 1982) (Fig. 4.4). Response to vernalization may be obligate or quantitative; in some long-day species, exposure to short days may completely substitute for vernalization, or both may be required. Generally, plants originating from higher latitudes or altitudes require longer periods of vernalization than those from lower latitudes and altitudes (e.g. Cooper, 1963; Thomas, 1979). No vernalization requirements have yet been found for tropical plants although temperatures well below the optimum for growth can promote floral initiation in some tropical legumes.

A range of about 60 days occurs in the flowering

time of cultivars of the annual, subterranean clover, in southern Australia (Fig. 4.3). This is the result of some differences in juvenility, the control of floral initiation by a variable (slight to strong) vernalization requirement, moderate response to photoperiod and, once these are met, some process which depends on high temperatures, particularly in the period from initiation of flowering (flower development) (Devitt, Quinlivan & Francis, 1978; Aitken, 1985). Flowering of *T. glomeratum* (Woodward & Morley, 1974) and annual medics (*Medicago* spp.) are similarly controlled: response to vernalization and long days varies among medic selections (Fig. 4.4) but generally vernalization and long days are additive, and they are able to substitute for each other in accelerating flowering.

Using temperature and daylength as driving variables, Clarkson & Russell (1979) were able to predict the flowering time of *M. scutellata* and *M. truncatula*. A quadratic function gave daily fractional increments of development between two stages in the flowering process:

$$a_1(T - a_0) + a_2(T - a_0)^2 = 1 \qquad (4.3)$$

where a_0, a_1, a_2 are regression coefficients and $T°C$ is the mean daily temperature. Development proceeded most rapidly at about 26°C. The prediction of flowering in *M. truncatula* was improved by including a cold or vernalization requirement; this was described by a negative hyperbola, i.e. the daily fractional increment of vernalization was greatest at a minimum temperature of −4°C and negligible at 10°C. Thus, flowering is most rapid if the plant is exposed to *both* low temperatures (as described by a vernalization requirement which is met most rapidly at −4°C) and subsequently temperatures of *c*. 26°C.

Fig. 4.4. The effect of different periods of vernalization of germinated seed on time to flowering of three annual medics grown under two different photoperiods, namely natural (Warwick, 28°S, daylight 10.7–13.2 h (▲) and 18 h (●)). Data in (a), (b) and (c) are for *Medicago scutellata*, *M. polymorpha* and *M. truncatula* respectively. The number of weeks date from midwinter (23 July). (Adapted from Clarkson & Russell, 1975.)

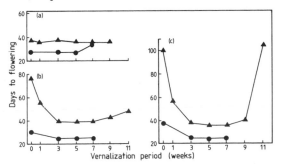

The photoperiod may interact with temperature to control floral initiation. Plants may thus have an absolute photoperiod requirement at one temperature but a quantitative or no response to photoperiod at other temperatures. In the field, seasonal changes in daylength and temperature (although slightly out of phase) provide appropriate cues for a diversity of flowering responses even in closely-related varieties; this can lead to errors in predicting varietal adaptation to new environments even at the same latitude. Low temperatures may extend the critical photoperiod required for the flowering of SDPs or shorten that required for the flowering of LDPs. This can be of adaptive significance. Changes of about 30 min in the critical photoperiod due to differences in the mean temperature of about 6°C between sites can mean the difference between non-flowering and flowering, and hence seed production, in some short-day species grown in the low-latitude tropics (Ison & Humphreys, 1984*b*). In the field in New Zealand and Australia the most significant effect of photoperiod and temperature interactions on flowering (e.g. Ketellapper, 1969) is the initiation of flowering in LDPs during the short days of autumn, winter and early spring. This is particularly evident in the earlier, more vigorous flowering of plants of Mediterranean origin compared with that of plants from higher latitudes.

Light intensity (quantum flux density) and light quality may affect flowering. Tillering, and thus potential flowering sites, is controlled by light quality and quantum flux density (Section 3.2). Low levels of light restrict photosynthesis and this restriction may delay floral initiation and the appearance of the inflorescence and reduce the number of flowers. Ryle (1967) has shown that the rate of production of leaf primordia and tiller fertility are reduced and the time of floral initiation is delayed in shaded plants of *Lolium* and *Festuca* (Fig. 4.5). This applies to many herbage plants and has implications for their use as intercrops and under plantation crops such as coconuts, where low light intensity may limit flowering and seed production and consequently their persistence (Section 8.3).

The amount, seasonal distribution and variability of rainfall delineates zones in which plants may grow. Adaptation to periods of moisture stress has occurred in five ways:

(i) The timing of phenological events to match the environment (e.g. Cameron, 1967). In this way many annuals escape moisture stress in the form of seed.

(ii) Opportunistic reproduction whenever soil moisture and temperature permit growth due to relaxation of daylength and temperature controls on flowering as seen in some Australian ecotypes of *Danthonia caespitosa* (Hodgkinson & Quinn, 1978).

(iii) Avoidance or tolerance, especially by perennials. Avoidance may occur through increased rooting depth and through sensitive stomatal control, leaf movements or reductions in leaf area, as in the tropical legume Siratro. Tolerance to water stress of e.g. −13 MPa (Ludlow, 1980) enables survival for flowering rather than affecting the flowering process directly.

(iv) Inhibition of flowering by short-term water deficits (Fig. 4.6).

(v) Promotion of flowering by short-term water deficits. There is frequent reference to this in the literature; a drying cycle induced by withdrawal of irrigation is sometimes used as a managerial tool to promote flowering and seed production in indeterminate tropical legumes such as *Macroptilium atropurpureum* cv. Siratro.

Fig. 4.5. The effect of shading on (a) the rate of accumulation of vegetative primordia and (b) the proportion of fertile shoots in ryegrass and fescue. ■, Grown in the United Kingdom in natural daylight; ▲, grown in 25 per cent light. (From Ryle, 1967.)

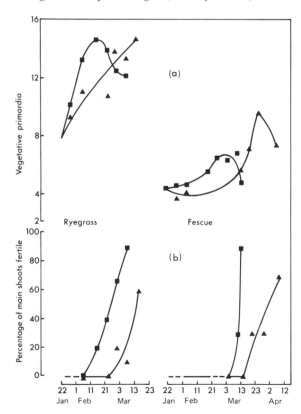

Water deficits do not have as much effect as photoperiod on the flowering of species but they may have more marked effects on later development. For instance, once flowering has begun, moisture stress may shorten the duration of flowering, as in medics (Fig. 4.6) and subterranean clover (Andrews, Collins & Stern, 1977; Taylor & Palmer, 1979).

Low nutrient status may retard rates of leaf appearance as well as retarding floral initiation. High levels of nitrogen accelerate flowering in herbage grasses by 10–20 days but this response varies between species and between cultivars. The role of nutrition on flowering and seed yield are considered more fully in Hebblethwaite (1980) and Humphreys (1979) and management implications in Section 4.5.6.

4.3.2
Autonomous flowering

This describes those plants which are day-neutral (DN) with respect to photoperiod and which do not require vernalization for floral initiation. Under such circumstances a degree-day or heat-sum model frequently predicts key developmental stages such as flowering because temperature determines the time of floral initiation via its effect on the rate of leaf production (e.g. Ong, 1983). This may also be true of species

Fig. 4.6. Linear trend of the effect of increasing water stress from M1 (no stress) to M4 (severe stress) on the time interval in days between various growth stages in *M. truncatula* cv. Jemalong. A, Planting to first flower; B, planting to first immature pod; (C), planting to first mature pod; (D), length of flowering; (E), planting to death. Regression coefficients (r^2): A, 0.95; B, 0.86; C, 0.81; D, 0.74; E, 0.93. (Adapted from Clarkson & Russell, 1976.)

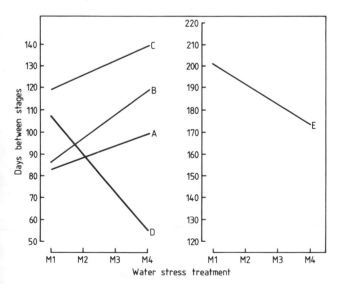

growing within their critical range of photoperiod, as for example lucerne regrowth over summer, or of plant development following exposure to low temperatures which satisfy a requirement for vernalization or low temperatures before flowering.

4.3.3
Development of the inflorescence

The development of the inflorescence extends from floral initiation to the appearance of the flower or inflorescence: i.e. to anthesis or blooming. The duration of development of the inflorescence up to the appearance of the flower(s) depends on (i) photoperiod; (ii) temperature and (iii) growth rate of the whole plant.

Vince-Prue (1975) recognizes four major groups of plants according, in part, to their response to photoperiod in terms of their inflorescence development:

  (i) plants with a photoperiodic requirement for floral initiation but which are DN for flower development;
 (ii) plants which are DN for floral initiation but which have a photoperiodic requirement for flower development;
(iii) plants with the same photoperiodic requirements for floral initiation and further development; and
(iv) plants with different photoperiodic requirements for initiation and further flower development.

The length of the phase from initiation to flower appearance is generally related negatively to temperature. This makes this phase of greater significance in cool environments; there are examples of initiation of grasses occurring over a seven-month period of low temperature whereas flower appearance is reduced to less than two months in spring (Evans, 1964). At supra-optimal temperatures, this phase may be delayed; for instance in Cook stylo the phase of inflorescence development decreased from 64 to 18 days with increasing temperature (T) over the range 18–27°C, so that the time from seedling emergence to flowering (y) was described by the function (Ison & Humphreys, 1984b):

$$y = 438.2 - 22.3T + 0.379T^2 \qquad (4.4)$$

This equation predicts a minimum time to flowering at 29°C, which is comparable with a minimum for lucerne at about 30°C (Field *et al.*, 1976). Above this temperature flower development may be inhibited.

Direct effects of temperature on inflorescence development are confounded with effects of temperature on the growth rate, which in turn affects the rate of inflorescence development. Temperature and radiation affect rates of node appearance and it is frequently difficult to separate this effect from direct effects of the environment on the flowering process. For example, a day-night temperature regime of 35/30°C delays flowering of two lucerne cultivars and it also delays the appearance of nodes (Pearson & Hunt, 1972*a, b*).

4.4
Fertilization and seed formation

4.4.1
Breeding systems

Herbage plants reproduce sexually or asexually; often one species will reproduce using both systems depending on the weather. Breeding systems, which determine the amount of genetic recombination and thus variability in grasslands, are summarized in Table 4.1. The morphology of flowers differs between legumes and grasses (Fig. 4.7); legumes are predominantly self- or cross-fertilizing whereas agamospermy is common among grasses.

Breeding systems are modified by environment, particularly photoperiod and temperature (e.g. Evans, 1964; Ison & Hopkinson, 1985).

Anthesis of sexually-reproducing species begins when anthers and stigmas are exposed to pollinating agents (air movement, insects) when the flower opens

Table 4.1. *Breeding systems found in pasture grasses and legumes*

System and form	Examples
Sexual (amphimixis)	
I Self-fertilization	
(i) Cleistogamy (pollen fertilizes home ovary before the flower opens or flower does not open)	*Danthonia* spp., *Stylosanthes* spp., *Trifolium subterraneum*
(ii) Pollen shed before the flower opens	*Medicago polymorpha*
II Cross-fertilization	
(i) Differences in size, location or function of the anthers and style	*Oxalis pes-capre*
(ii) Faster growth rate of 'foreign' pollen down the style	
(iii) Dichogamy (the anthers and stigmas become functional at different times: (a) protandry (the anthers shed pollen before the stigmas are receptive); (b) protogyny (the stigmas are receptive before the pollen is shed)	
(iv) Self-incompatibility (genetically controlled)	*Trifolium repens, Lotus corniculatus*
(v) Monoecism (male and female flowers separate on the same plant)	*Zea mays*
(vi) Dioecism (male and female inflorescences borne on separate plants)	*Spinifex hirsutus*
(vii) Gynodioecious or androdioecious (individuals are either hermaphrodite or female or male respectively)	*Arundo richardii*
(viii) Gynomonoecious and andromonoecious (hermaphrodite and female or male flowers borne separately on the same individual)	*Buchloe dactyloides*
Asexual (apomixis)	
I Agamospermy (seeds are produced)	
(i) Adventitious embryony (one or more adventitious embryos develop in the ovule directly from somatic cells of the nucellus or integument: (a) the endosperm develops without a need for fertilization; (b) fertilization is necessary to 'trigger' endosperm formation)	*Poa pratensis*
(ii) Gametophytic apomixis (includes all forms of apomixis morphologically similar to the sexual process, but meiosis is avoided or is modified so that the embryo sac is diploid and the embryo develops without fertilization). There are two forms: diplospory and apospory, with the latter being the most common	Many tropical grasses including *Bothriochloa, Dicanthium, Paspalum, Chloris, Poa, Panicum*
II Vegetative reproduction (no seeds are produced)	
e.g. tillers, stolons, rhizomes, cuttings, runners	Some ecotypes of *Cynodon, Pennisetum clandestinum*

or because these organs protrude from a closed flower. Anthesis ends when the same organs are no longer available to the agents of pollination. In cleistogamous grasses or legumes (those where pollination and fertilization occur within closed florets), there is, by definition, no anthesis.

Outcrossing rates in most cleistogamous herbage legumes range from 0.01 to 10 per cent. Subterranean clover has approximately 0.1 per cent outcrossing (Marshall & Broue, 1973). This is sufficient to produce 10–20 hybrids per m² per year. Small and varying degrees of outcrossing result in the evolution of genetic diversity even in sown pastures. In southern Australia new strains of subterranean clover develop in mixed grasslands at a rate of 1.5 per cent of the total population per year (Reed & Cocks, 1982). Consequently there is increasing interest in sowing complex mixtures of screened genotypes to take advantage of this potential to adapt to a wide range of micro-environments.

Outcrossing does, however, cause problems for seed producers because of genetic shift or genetic contamination of cultivars. Against this, outcrossing provides a potential to breed facultatively apomictic grasses such as *Bothriochloa intermedia* and *Cenchrus ciliaris*.

Further examples, and the basis of herbage plant breeding, are reviewed in McIvor & Bray (1983). The

potential exists to generate new gene combinations by techniques such as embryo culture, cell selection, somatic hybridization and recombinant DNA technology. Somaclonal variation exists in lucerne, clovers, *Lolium*, *Festuca* and *Panicum*; cell lines of lucerne and white clover are resistant to some herbicides (Scowcroft *et al.*, 1982). Vasil (1986) reviews progress in the application of 'biotechnology' to herbage grasses.

4.4.2
Anthesis and fertilization

Anthesis is affected by the environment. Some generalizations are: (i) low temperatures and frost may inhibit anthesis or induce pollen sterility, as in kikuyu; (ii) the pattern of timing of anthesis during the day is temperature-dependent and varies between and within species; (iii) it is difficult to isolate the effects of light, total radiant energy, relative humidity and ambient temperature in much research and in the common observations that anthesis is prolonged in some legumes on mild overcast days whilst florets of grasses are said to stay closed on 'wet, cold, dull days'; (iv) weather, particularly temperature conditions, on the day prior to anthesis or during the night may be more critical than that on the day of anthesis; (v) high temperatures may inhibit anthesis through desiccation, style injury or arrested anther development; (vi) high wind velocity, e.g. 30 km per h may inhibit anthesis in some but not all cross-pollinating species (e.g. Hill, 1980).

Pollination describes those processes occurring from the time of anther dehiscence until the pollen tube reaches the embryo sac in the ovary. Failure of

Fig. 4.7. Generalized flower structures. (a) Grasses showing a single floret closed (i), a single floret open (ii), floral parts (iii) and a spikelet (iv). (b) A dicotyledon, e.g. legume, flower. (Adapted from Jackson & Jacobs, 1985.)

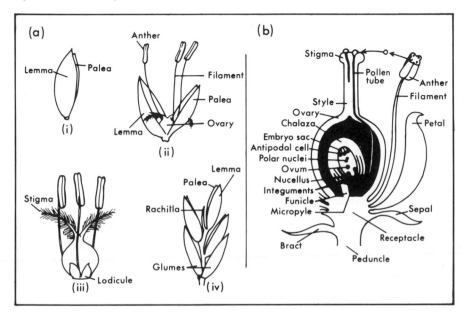

pollination can be a major cause of seed crop failure in cross-pollinating species such as lucerne. Insect pollinators may not exist or they may be killed by pesticides. Empirical research enables some prediction of the desirable bee hive density necessary for lucerne pollination by the honey bee (*Apis mellifera* L.) but it is better to have a greater understanding of insect behaviour and crop phenology so as to devise appropriate management strategies. Modelling may assist: the relationship between the number of florets visited by the bumblebee (*Bombus fervidus*) and seed set has been derived in red clover (*Trifolium pratense*) using a probabilistic model for the expected number of ovules fertilized (Plowright & Hartling, 1981).

In some species pollination depends on overcoming physical barriers. For example, in lucerne tripping of flowers (rupturing wing and keel petals to release a restrained staminal column so that it springs against the standard petal rupturing the stigmatic membrane) is essential to allow cross-fertilization. Tripping may be automatic in 1–14 per cent of flowers at 25°C and positively related to temperature over the range 20–40°C.

Fertilization is the fusion of the male nucleus with the ovum. In grassland plants, as in all angiosperms, embryogenesis commences with double fertilization which gives rise to the zygote and a triploid cell from which endosperm tissue develops. These events occur in the embryo sac embedded in maternal tissue, which is described in more detail by Dure (1975).

Seed set, a term used frequently in grassland seed production studies, is best considered as the commencement of cell division following successful fertilization. Unless cell division occurs the ovules shrivel and die: effective pollination and fertilization do not guarantee seed formation.

Seed development follows seed set; cell division continues until the embryo has a fully-developed scutellum, a stem apex and root initials. Finally, desiccation occurs as the ovule loses water to the surrounding environment and the seed coat tissue sclerifies and dies, thereby encasing the endosperm/embryo in a protective layer. Seed does not have to develop fully to be viable; seeds of *Pennisetum americanum* have normal vigour if they reach 50 per cent development (Fussel & Pearson, 1980).

In legumes, the proportion of 'hard' seed is determined by the development of cells and tissues at the strophiole (Mott, 1979) and the hilum (Hyde, 1954) as the seed progresses towards physiological maturity. Hardseededness is determined by genotype, the environment during seed development and environmental factors controlling the breakdown of hardseededness after the seed is dispersed or sown

(Section 2.2). In *Stylosanthes hamata* cv. Verano, the temperature during seed development is the most significant factor in determining the percentage of hard seed, more so than variations in the availability of moisture in the soil or of light. The percentage hardseededness is positively related to the temperature of provenance indicating that seed developing at cool sites, late in the season, or when temperatures are lower than normal, would be expected to contain a lower proportion of hard seed (Argel & Humphreys, 1983). This has implications for purchased seed lots. It suggests that seed should be selected from hot locations to maintain high levels of hardseededness and thus longevity of seed reserves in the soil. In subterranean clover hardseededness is correlated with seed size: the largest seeds soften first (Taylor & Palmer, 1979), as do those seeds which develop from the latest flower within the inflorescence (Salisbury & Halloran, 1983). Further aspects of hardseededness are discussed in Chapter 2.

4.4.3
Seed production

Flowering occurs either simultaneously or staggered at sites within the plant and among plants in a grassland. The result is a population of potential seed-producing sites in varying stages of reproductive development. When managing and locating commercial seed production it is common to refer to the components of seed yield which may be manipulated advantageously given appropriate understanding of these components and their interactions. It is not possible to deal with the many aspects of commercial seed production here and readers are referred to recent reviews for tropical grasslands (Humphreys, 1979) and temperate pasture seed crops (Hebblethwaite, 1980; Lancashire, 1980). One or more of the components of seed yield may also determine ecological success; this is related to the reproductive capacity of the individual plant (Eqn (4.1)). For a population of plants, the viable seed yield per unit area (Y) is given by:

$$Y = t \times lf \times f_t \times sp \times fl \times f_{fl} \times s_{fl} \times V_s \qquad (4.5)$$

where t is the number of tillers or bud sites per unit area, lf is the percentage of these surviving to flowering, f_t is the percentage of surviving sites which are fertile (the product of these equals the inflorescence density); sp is the number of floral units (e.g. spikelets) per inflorescence, fl is the number of flowers differentiated per floral unit, i.e. spikelets; f_{fl} is the percentage of fertile flowers (seed-bearing), s_{fl} is the number of seeds formed per flower and V_s is the percentage of viable seeds. The component of Eqn (4.5) which contributes most to the seed yield varies with the species, location (environment) and management (e.g. plant spacing); the inflorescence density has the greatest effect on the seed yield of perennial grasses (e.g. Langer, 1980).

For commercial harvest, the factors S_w (individual seed weight) and S_h (the percentage of seed actually harvested) should be added although increased seed size usually imparts a competitive advantage to the germinating seedling by way of increased vigour.

A similar seed yield may be achieved by diverse pathways; compensation between yield components may limit the variation obtainable in the final yield. This may be seen in Fig. 4.8, which shows the response of Y and some components to plant density. Manipulation of either the duration or the rate of one of the development phases which contribute ultimately to seed yield is of major interest to plant breeders, agronomists and managers. Charles-Edwards (1982)

related seed yield to the duration of the reproductive growth phase by:

$$W_H = \eta_H \varepsilon \overline{Q} \overline{S} \Delta t + b \qquad (4.6)$$

where W_H is the dry weight of seed harvested, η_H is the proportion of new dry matter partitioned to the seed, ε is the efficiency of light utilization, $\overline{Q}$ is the average proportion of the incident light flux density intercepted by a plant or crop over an extended period of time, $\overline{S}$ is the integral of the average daily light incident on the crop during reproductive growth, Δt is the duration of the reproductive growth phase and b is a constant. Eqn (4.6) does not always apply; the inflorescence density, and ultimately the seed yield may be determined by the rate and/or the duration of inflorescence production. For example, the seed yield of subterranean clover in Australia is: (i) positively correlated with the rate of inflorescence production (1.1–2.4 per dm^2 per day) but is not correlated with the time or duration (32–48 days) of flowering (Francis & Gladstones, 1974); (ii) increased by defoliation, which increases the rate of inflorescence production (Collins, 1978); (iii) reduced by shading, which reduces the rate but does not prolong the duration of flowering (Collins, Rossiter & Ramos Monreal, 1978); (iv) reduced by severe water stress, which reduces the rate of flowering and/or its duration (Wolfe, 1981); (v) reduced by high light during seed development in species which bury their seeds (the seed weight per burr is negatively and linearly related to the logarithm of the light intensity (Taylor, 1979)); and (vi) promoted by increasing the day/night temperature regime up to 24/19°C, which causes an increase in the rate of inflorescence production and a reduction in the duration (Taylor & Palmer, 1979). At a day/night temperature regime of 24/19°C the seed yield was reduced due to a decrease in seed size but not in seed number.

The main effects of temperature on seed yield components can be summarized as follows:
(i) The final yield (Y) may be determined more by temperature effects on tillering than on any reproductive stages. For example, in the tropical grass *Pennisetum americanum*, Y is maximal when low day/night temperatures (21/16°C) increase basal tillering even though this temperature is unfavourable to spikelet fertility and inflorescence length (Fussell, Pearson & Norman, 1980).
(ii) The optimum temperature for developmental events frequently declines during reproductive development; in Cook stylo growth and the time to flower appearance was optimal at 27°C whereas inflorescence production and floret number per inflorescence were highest at 24°C (Ison & Humphreys, 1984b). Akpan & Bean (1977) found that

Fig. 4.8. Effect of plant density on seed yield and its components in *Stylosanthes humilis*. (From Shelton & Humphreys, 1971.)

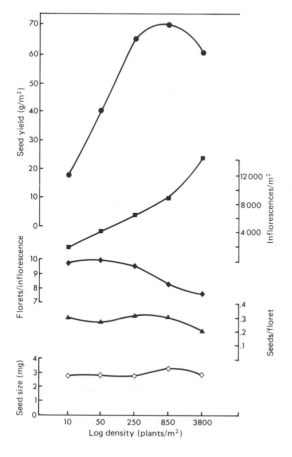

20°C was the optimum temperature for floret fertilization in *Lolium* and *Festuca*; lower temperatures (*c.* 15°C) were required for seed formation (producing larger ovaries and larger seeds) and higher temperatures (*c.* 23°C) during seed development were most favourable for subsequent germination vigour.

(iii) In most legumes the inflorescence density is reduced above a mean temperature of 30°C. In temperate pasture grasses spikelet initiation probably increases with temperature to a maximum at 20–25°C and the duration of spikelet production is shortest at 25–30°C (Brooking, 1979).

(iv) High temperatures, especially at night, frequently reduce floret and seed numbers and seed weight by shortening the duration of inflorescence and later seed development. It is likely that a reduction in seed size in these conditions is due to a reduced cell size rather than a lower cell number (Wardlaw, 1979).

4.5
Implications for grassland growth and management

Plants sown in grasslands fall broadly into five categories with respect to their flowering and seed-producing characteristics:

(i) Temperate LDPs which grow in areas of reliable growing season and show synchronous flowering in spring and early summer.

(ii) Mediterranean LDPs in which flowering and subsequent seed production may be curtailed by high temperatures and/or summer drought.

(iii) Tropical legumes and grasses which are responsive to photoperiod (usually but not invariably SDPs) for flowering and which are often well synchronized but where flowering and seed production may be curtailed by low temperatures and/or winter (dry season) drought.

(iv) Tropical legumes, usually indeterminate, which are not under strict photoperiodic control for flowering; these characteristically show several cycles of often bimodal flowering each year.

(v) A large group of tropical grasses whose flowering is poorly synchronized and occurs whenever conditions are favourable for growth. This usually means flowering occurs in the wet season whether species are LDP, SDP or DN.

The implications of flowering for growth and management vary between these broad groups. Most is known of the first group. Reproduction in this group is confined to a short period in spring although in 'stable' plant communities there may be a tendency towards decreased synchrony of reproduction (Newell & Tramer, 1978). This may be desirable for redistribution of some of the high spring yields.

4.5.1
Flowering and growth rate

Reproductive development may affect the net efficiency of utilization of radiation and thus grassland growth directly as well as by the more obvious effects of reproductive tillers carrying young expanding leaves to a more favourable light environment at the top of the canopy (e.g. Woledge, 1979). In determinate grassland species the change from vegetative to reproductive state causes no more leaves to be initiated on that tiller. Thus the rate of leaf appearance varies seasonally (after the effects of temperature and radiation are taken into account) (Davies & Thomas, 1983). In compensation, the rate of leaf elongation and the growth rate of the remaining leaves increases (Williams & Biddiscombe, 1965). The growth rate of the whole tiller then increases, as in ryegrass (Davies, 1971) due to both an increase in the unit shoot rate (u, where $u = \mathrm{d}w/\mathrm{d}t \times 1/W_s$ and W_s is the weight of the shoot system) and a decrease in the proportion of photosynthate used in root production as well as an increase in the rate of net photosynthesis (Troughton, 1977). Apparent ε, the efficiency of utilization of radiation (shoots only), increases by 40 per cent (Anslow & Green, 1967) to as much as 400 per cent (Field, 1980; Section 3.3) during the elongation and emergence of the inflorescence. Stem elongation is not the only explanation of the enhanced photosynthesis as the photosynthetic capacity of the leaves in the flowering sward is not reduced by moderate shading (Woledge, 1978). Parsons & Robson (1982) have related the physiological basis of the superiority of the growth rate of reproductive ryegrass over the growth rate of vegetative swards to: (i) the extension by reproductive plants of leaves of high area per unit weight despite low temperature; (ii) the mobilization of carbohydrates, particularly fructosans, for growth during this period of low photosynthesis; (iii) the investment of carbon by the reproductive plant in tissues most effective either directly (leaf tissue) or indirectly (elongated stem) in achieving and maintaining a canopy of high photosynthetic potential; and (iv) the longer retention of this carbon (in flag leaves, elongated stems and ears) on the plant than of the retention of carbon in the leaves of vegetative swards.

Growth is also reduced in the flowering grassland, due to inhibition of tillering (Davies, 1971). Tillering may, however, increase during seed maturation if conditions are favourable for growth; yields of shoot dry weight from herbage grasses are commonly correlated with the number of elongating or heading tillers (e.g. Inosaka *et al.*, 1978). In contrast, inflorescence development in indeterminate species probably slows both the rate of leaf production and shoot growth by diverting photosynthate from the lamina to the inflorescence.

4.5.2
Flowering and quality

Flowering and the associated elongation of stems and production of reproductive parts causes a decrease in the leaf/stem ratio and root/shoot ratio. There is also a reduction in growth and often a net loss of root hairs and nodules (Fig. 3.1). These result in a decline in quality and an increase in susceptibility to deterioration, e.g. leaf fall and spoilage. The bulk density of herbage also declines. Changes in quality and their effects on animal intake are discussed in Chapter 6.

4.5.3
Selection of cultivars

There are few direct ways of managing grasslands for controlled flowering and seed production within particular environments. The available management strategies include: (i) cultivar selection (addition of one or more genotypes to a seed pool); (ii) selection of sowing time to allow appropriate environmental control of development; (iii) defoliation management; (iv) strategic application of fertilizer, especially nitrogen; and (v) where feasible, irrigation. Variation in the flowering time of different ecotypes is a highly useful adaptive mechanism in natural populations and is frequently the basis for commercial cultivar development and release. Flowering time is usually highly correlated with the length of the potential growing season in the original habitat of an ecotype (e.g. Cooper & McWilliam (1966) for *Phalaris aquatica*). However, it may vary because of one or more of the controlling factors discussed earlier (Section 4.3). A particularly good example is the photoperiod-sensitive (SDP) annual tropical legume *Stylosanthes humilis* (Edye & Grof, 1983). Within naturalized Australian populations of this species there exists a range of maturity types which differ in the critical photoperiod they require for flowering. 'Early' flowering types (those with a longer critical photoperiod) come from higher latitudes and drier sites with a shorter growing season whilst 'late' maturing types come from frost-free low latitude and wet sites. The 'early' maturing lines dominate at drier sites because they are able to flower and set seed before the onset of moisture stress, unlike the 'later' types; at the wetter sites the more vigorous, higher dry-matter-yielding, 'late' flowering plants out-compete the smaller 'earlier' flowering

types when sown at a common date (Fig. 4.9). Such differences have been used commercially, e.g. with ranges of annual *Medicago* spp. and a range of sub terranean clover cultivars, two examples of which are shown in Fig. 4.3.

In areas which have an extremely variable spring climate, early and prolonged flowering and maturation give the potential for cultivars to maximize herbage and seed yield. Sowing mixtures of seed of ecotypes of varying time to maturity may enable adaptation to sites within individual paddocks which differ in their potential growing season; rapid changes in the dominance of one genotype over another in mixed genotype populations of annuals are possible in these circumstances.

4.5.4
Sowing time

Species having a long juvenile phase or requiring vernalization will not flower and return seed to the soil seed bank in the year of sowing if they are sown at the wrong time. If photoperiod-sensitive species are sown close to their critical photoperiod then leaf initiation ceases before many potential reproductive sites can be generated; this effect can be gauged from Fig. 4.3 for, particularly, *Trifolium subterraneum* cv. Dwalganup. This is one reason why temperate and Mediterranean species requiring long days for flowering are usually sown in the autumn.

Fig. 4.9. Fitted growth curves for the forage components (leaves and stems) and reproductive components of three ecotypes of *S. humilis* sown at the same time of year. (——), Late-flowering ecotype; (– – –) mid-flowering ecotype; (· · · ·), early-flowering ecotype. (From Fisher, Charles-Edwards & Campbell, 1980.)

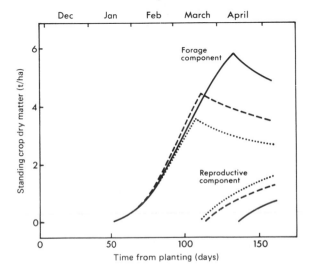

4.5.5
Management by defoliation

Given that cultivars suitable for the environment and farming system have been selected and that they are sown at a time suited to their purpose, then the most important strategy for managing return of seed to the seed pool (Fig. 2.2) is defoliation. Defoliation encompasses grazing and cutting (or the lack of them) and the timing of these with respect to cultivar phenology or variations in phenology between cultivars or species making up the grassland.

Defoliation is usually carried out prior to the appearance of flowers and it may have a number of effects on flowering and seed yield. Some management strategies are:

(i) Defoliation to produce more bud sites by increasing axillary branching or the density of tillers; this is best carried out before extension growth carries the floral apex above the height of defoliation.

(ii) Defoliation prior to flowering to reduce shading (increase the light), thereby increasing the potential bud site fertility (e.g. Rossiter, 1961).

(iii) Severe defoliation during early growth to delay floral initiation, e.g. in subterranean clover cv. Mt Barker floral initiation was delayed (by 30 days), as was the subsequent rate of leaf appearance (Collins & Atiken, 1970). The delay of floral initiation varies between ecotypes. In a later study a minimum increase in seed yield of 30 per cent was obtained in swards of each of three cultivars cut weekly until flowering via a slight delay in floral initiation, no delay in flower appearance, increases in the rate and number of inflorescences produced and increased burr burial (Collins, 1978).

(iv) Defoliation to increase the hard seed component of legume seeds and reduce the rate of hard seed breakdown (e.g. Collins, 1978). Defoliation of subterranean clover probably increases hard-seededness by increased burr burial.

(v) Defoliation to manipulate the competitive relations between cultivars of a species (e.g. Collins *et al.*, 1983) or between species (e.g. Smith & Crespo, 1979).

(vi) Defoliation to maximize the spring yield for hay or silage (e.g. Swift & Edwards, 1983). In paddocks cut for hay or silage or grown for seed this results over time in genetic shifts towards early-flowering populations that are able to set seed before cutting (e.g. Clements, Hayward & Byth, 1983).

(vii) Defoliation to improve the synchrony of tiller development and head emergence in grass and some legume seed crops (Ison & Hopkinson, 1985).

(viii) Exclusion of stock to allow flowering and seed set. This, for example, increases the persistence of Siratro-based grasslands in Thailand.

Where defoliation occurs on reproductive swards, regrowth is reduced due to lower tiller numbers, the cutting off of apices of reproductive tillers, low stubble weight, lower carbohydrate reserves and reduced root weight.

4.5.6
Fertilizer application

Here we consider the role of nitrogen fertilizer as a management tool for grassland seed production; mineral nutrition is discussed generally in Chapter 5. Extensive work on the nitrogen fertilization of perennial ryegrass seed crops (Hebblethwaite, Wright & Noble, 1980) has shown that the timing and quantity of nitrogen applications affects various components of the seed yield. Variation in the number of seeds produced per unit area accounts for 60–98 per cent of the variation in yield (Hampton & Hebblethwaite, 1983). The number of seeds is a product of the number of fertile tillers per unit area, the number of spikelets per fertile tiller and the number of seeds per spikelet. The practical implications from the work of Hebblethwaite, Wright & Noble (1980) and from other work are:

(i) The application of nitrogen before, or at the latest at floral initiation increases the density of the

Fig. 4.10. Accumulation of toxin in grassland harvested at weekly intervals during spring and early summer in Western Australia and reproductive phases of annual ryegrass (*Lolium rigidum*) at corresponding times. (Adapted from Styles & Bird, 1983.)

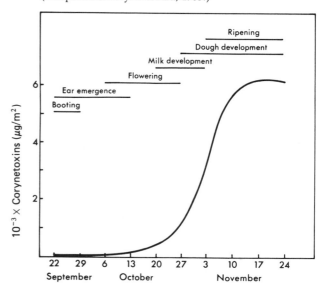

fertile tillers, but application at spring closing (i.e. after floral initiation) reduces the density of fertile tillers even though the total tiller density is increased. The optimum density is 2000–4000 fertile tillers per m².

(ii) Late application of nitrogen increases the likelihood that tillers will fall over onto the ground (lodge) although in the United Kingdom many crops are lodged before anthesis. Under these conditions the number of seeds set per spikelet is critical.

(iii) The optimum quantity of nitrogen to apply varies with the soil type and residual soil nitrogen.

(iv) The application of nitrogen may have opposing effects on the rate, and thus the duration, of different developmental phases. In *Poa annua* high levels of nitrogen reduced the period from sowing to inflorescence emergence from 68 to 43 days but increased the period to subsequent seed maturity from 22 to 29 days (Ong, Marshall & Sagar, 1978). Humphreys (1979, 1981) reviews this subject in greater detail.

4.5.7

Diseases and pests associated with flowering

Diseases and pests associated with reproductive development are also of managerial interest. In annual ryegrass (*Lolium rigidum*) galls induced by the nematode *Anguina agrostis* may become colonized by the bacterium *Corynebacterium rathayi* to produce toxins during reproductive development (Fig. 4.10). The toxins may result in high animal mortalities (Stynes & Bird, 1983). Ergot (*Claviceps paspali*) infection

of seed heads of *Paspalum dilatatum* may have similar effects (Hampton, 1984). Both nematodes and ergot reduce yields of viable seed, as do other organisms, e.g. the fungal pathogens *Botrytis cineria* and anthracnose in legume seed crops (Table 3.4), which do not create health problems for the grazing animals. Seed-eating beetles (*Curculionidae*) and bud worm (*Stegasta bosqueela*) reduce tropical legume seed production whilst seed-sucking leaf hoppers (*Cicadellidae*) reduce tropical grass seed production (Lenné, Turner & Cameron, 1980). These diseases and pests may be minimized by (i) selecting cultivars which are not susceptible to infestation, e.g. *Lupinus angustifolius* cv. Uniwhite is resistant to *Phomopsis leptrostomiformis*, a fungus which produces toxins on lupin stubble in humid conditions and causes lupinosis, a liver disease in grazing livestock; (ii) grazing management which includes mixing of the diet with, for example, roughage; (iii) controlled burning; (iv) biological control where warranted; and (v) chemical control where economically and environmentally feasible.

4.6
Further reading

Barnard, C. (ed.) (1966). *Grasses and Grasslands*. London: Macmillan. Chapters by Evans on reproduction, McWilliam on cytogenetics and Latter on breeding of grassland plants.

Ison, R. L. & Humphreys, L. R. (1984). Reproductive physiology of *Stylosanthes*. In *The Biology and Agronomy of Stylosanthes*, ed. H. M. Stace & L. A. Edye, pp. 257–77. Sydney: Academic Press. A case study of flowering and seed formation.

5

Mineral nutrition

Sixteen elements (excluding carbon, hydrogen and oxygen) are essential for grassland growth. These are the macronutrients nitrogen, phosphorus, potassium, sulphur, calcium and magnesium and the micronutrients iron, manganese, zinc, copper, molybdenum, chlorine, boron, cobalt, selenium and sodium. In addition, iodine is required by the grazing animal.

These inorganic elements are distributed among four compartments within the grassland system (Fig. 5.1). They are replenished within the system by the application of fertilizer, the release of minerals through the breakdown of organic matter, the weathering of parent rock, run-on through lateral movement of drainage water, and in rainfall and through the fallout of particulate material. Elements are lost through the removal of animal or plant products, e.g. milk, beef, wool, hay, silage; through run-off and leaching below the root zone; and through volatilization to the atmosphere.

Because there are only these four compartments within the mineral cycle, it is relatively simple to describe the change with time in any one compartment, or the whole system, by equations such as:

$$\Delta P = f(F \rightarrow P) + f(S \rightarrow P) - f(P \rightarrow D) - f(P \rightarrow A) \quad (5.1)$$

where ΔP is the change in mineral content (kg element per ha per year) of the pasture compartment (shoots plus roots) and the fluxes (f) are the rates of flow of element from fertilizer (F) and soil (S) to the live pasture plant (P) and the losses from the live plant to grazing animals (A) and to dead plant material (including hay and silage) and non-plant organic material: the detritus food chain (D).

The cycle is fundamentally a self-sustaining circle wherein the bulk of the flow is from soil through plants, animals and the detritus chain back to inor-

ganic elements in the soil. That is, inputs balance outputs in the long term if grassland production is stable. In many situations this balance is maintained by the application of fertilizer. In some situations where long-term depletion of elements is not severe, it is unnecessary to intervene through the application of fertilizer. Three examples of these situations are:

(i) Alluvial soils which have high fertility, e.g. the alluvial lowlands of Indonesia. Here elements which are lost through product removal and run-off are replenished by run-on of nutrients from

Fig. 5.1. Dry matter and element cycling within the grassland system. The compartments are live pasture, P; animals (grazing herbivores), A; dead material, D; and available soil elements, S.

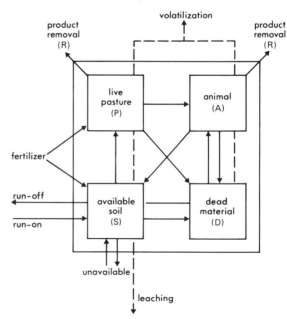

other, higher regions. These areas are usually used for crop, not grassland, production, and often neither crops nor grasslands respond to macronutrients.

(ii) Soils which have high amounts of organic matter and which are in cool temperate climates where rates of both mineral loss and grassland growth may be low due to low radiation and temperatures. In such regions plant uptake may be related closely to rates of mineralization from organic matter.

(iii) Extensive unfertilized savannah grasslands in which element removal by gathering and 'trucking out' of low numbers of steer beef is small, e.g. 10 kg N per ha per year. These losses can be replenished by the release of minerals through the weathering of parent rock.

A second generalization about the mineral cycle is that the balance of inputs and outputs, and the flux rates and compartment sizes within the system, differ between elements. More is known about the quantitative cycling of the macronutrients and molybdenum than of the other micronutrients; the commonly-occurring range of annual inputs, outputs and the distribution (pool sizes) of nitrogen, phosphorus and sulphur in an annual grassland is shown in Table 5.1.

There are seasonal fluctuations in the rate of flow of elements through various parts of the cycle. These fluctuations depend on management, e.g. the application of fertilizer, and on such factors as the time and location of return of dung, urine and plant residues, the intensity of grazing and the removal of product, and on the seasonality of the weather as it affects grassland growth and mineralization of elements from organic matter. The seasonal variation in mineral cycling is more pronounced for some components of the cycle than for others, as shown for the cycling of sulphur in Fig. 5.2.

In this chapter we consider first the major fluxes of elements within the grassland system (Fig. 5.1). These fluxes are to varying degrees amenable to management. Their management is often (and will become increasingly) based on a diagnosis of nutritional status by elemental analysis of the soil, of leaves, or of whole plants; this is discussed briefly in Section 5.7. In this chapter we conclude with some of the more important examples of management in relation to nutrition.

Table 5.1. *Ranges of annual outputs, inputs and distribution of nitrogen, phosphorus and sulphur in an annual grassland system, assuming carbon to nitrogen to phosphorus to sulphur ratios of 60 : 9 : 1 : 0.5*

Inputs, outputs and distribution	Nitrogen (kg/ha)	Phosphorus (kg/ha)	Sulphur (kg/ha)
Inputs and outputs			
← Leaching and run-off	13–63	0–0.5	1–20
→ Rain	3–10	0.4–0.5	1–20
→ Fertilizer		← Variable, as desired →	
→ Feed supplements	0–100	0–12	0–6
← Animal products	2–20	0.2–4	0.1–2
→ Nitrogen fixation	5–50	—	—
Distribution below ground			
Inert	—	6 000–12 000	100
Organic	2 000–5 000	1 500–3 500	750–1 750
Available mineral	1–10	1–20	1–10
Roots	20–80	2–8	2–15
Microbes	10–40		
Distribution above ground			
Plant tops	35–80	2–35	2–4
Herbivores	2–12	1–10	0.1–0.6[a]
Dung and urine	10–50	0.7–25	0.5–12.0
Carnivores	—	0.15–15	—

[a] 0.2–0.9 sheep/ha.

←, losses.

→, gains.

Source: Jones & Woodmansee (1979).

5.1
Fertilizing

5.1.1
Types of fertilizer

Russell (1973) and others give details of the chemistry of various elements applied as fertilizers to soil–plant systems. Here we present only a summary for some of the macroelements, nitrogen, phosphorus, potassium and sulphur, as a basis for later discussion of the agronomic implications of fertilizer use.

Nitrogen in fertilizers is bonded chemically as nitrate (NO_3^-), or as ammonium (NH_4^+) or as urea ($CO(NH_2)_2$). Nitrate is soluble in water and is therefore available for immediate uptake by the grass but it is also liable to removal by run-off or leaching below the root zone. Ammonium is readily dissociated from its carrier compound, whether it be chloride, sulphate or nitrate; it is also the product of bacterial (urease) breakdown of urea and the first product of bacterial degradation of soil organic matter. The release of nitrogen from organic matter, termed mineralization, is high under grasslands relative to that which occurs under forages and crops. For example, mineralization under grassland was 200–300 kg N per ha per year compared with 100–130 kg N per ha per year for barley in Sweden (Rosswall & Paustian, 1984). Ammonium, supplied from fertilizer or organic sources, can be adsorbed onto soil colloids. It is also volatilized, particularly at high temperatures, under dry conditions on alkaline soils. Volatilization can be related directly to the calcium content of soils: Lehr & van Wesemael (1961) found that loss of ammonium by volatilization increased from zero to 20 per cent as the content of calcium carbonate increased to 20 per cent of soil weight.

Ammonium is oxidized to nitrate by the bacteria *Nitrosomonas* and *Nitrobacter*. The net result of these oxidation reactions, collectively called nitrification, is:

$$NH_4^+ + 2O_2 \rightarrow NO_3^- + 2H^+ + H_2O \qquad (5.2)$$

Nitrification may occur rapidly (as much as 70 per cent within 5 days) after ammonium fertilizers are applied to pastures. Its consequence is soil acidification. Nitrification is slower in very dry soils at low soil temperatures (<7°C) and under low pH; it is restricted or inhibited in the absence of oxygen. Waterlogged grasslands commonly have surface water which is reasonably well aerated (7–12 mg O_2 per l) but in the waterlogged soil itself the level of oxygen is so low it is difficult to measure. In such soils denitrification – the reduction of nitrate to dinitrogen gas (N_2) and nitrous oxide (N_2O) – may result in loss of nitrate-based fertilizer as a gas to the atmosphere. Denitrification ranges between 0 and 7 per cent of fertilizer nitrogen applied to grasslands; it may account for as much as a 20 per cent loss of nitrogen from crops (Colbourn & Dowdell, 1984).

Phosphorus may be carried in water-soluble compounds, e.g. dihydrogen phosphate, $H_2PO_4^-$, in superphosphate and ammonium phosphates, or in slowly-soluble forms, e.g. apatite in rock phosphate. When a phosphate fertilizer is added to the soil, water enters the fertilizer granule and a solution is formed which is saturated with monocalcium phosphate and dicalcium phosphate dihydrate. This solution is extremely acidic (pH 1.5) in the immediate vicinity of the fertilizer granule. Thus, contact with the granule will kill rhizobia. Dihydrogen phosphate reacts rapidly with Ca^{2+} to form tricalcium phosphate, which is sparingly soluble and not acidic, or it is adsorbed onto clay colloids. Consequently, the concentration of phosphorus in solution in the soil is dilute (10–100 μM or 0.3–3 ppm: Mengel & Kirkby, 1979) and the mobility (and leaching) of fertilizer phosphorus is low. The bulk of the phosphorus exists either as rapidly-exchangeable (labile) precipitates or ions adsorbed

Fig. 5.2. Modelled behaviour of sulphur movement into various compartments of a temperate grass–legume–sheep grassland. (a) Change in sulphur content of fertilizer (F), pasture plants (P), dead material (D) and available pool in the soil (S) (——) with time after the application of fertilizer. (b) Change in sulphur contents as above except that two rates of transport of sulphur from fertilizer to available soil pool ($f(F \rightarrow S)$) are modelled: a high rate of transport (——) and approximately one-third that rate (---). Units are mCi (milliCurie), a measure of radioactivity from radioactive gypsum originally applied to the pasture. (From May, Till & Cumming, 1972.)

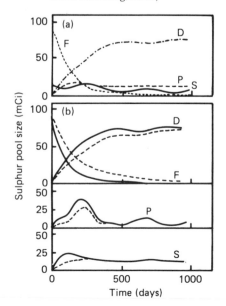

onto clay particles, or as non-labile ('unavailable') compounds.

Potassium is most commonly applied to grasslands in cationic form as potassium chloride, potassium sulphate or potassium nitrate. These are soluble in water. High losses due to leaching occur only on sandy soils, organic soils and soils in which the main clay mineral is kaolinite. This is because much of the K^+ added as a fertilizer is either adsorbed onto the surface of clay particles or is strongly adsorbed into the inter-layers of clay lattices, by potassium fixation. Potassium adsorbed in exchangeable positions on clay or organic matter is of later significance as a source of potassium for uptake by grasses; fixed potassium is unavailable. The abilities of different clay minerals to fix potassium are in the order vermiculite > illite > montmorillonite.

Inorganic sulphur is carried as sulphate (SO_4^{2-}) in fertilizers, e.g. superphosphate and gypsum. Sulphate is soluble and therefore liable to leaching: sulphur deficiencies are most common in grasslands in high rainfall zones whereas sulphur accumulates in the topsoil in dry climates. Where there is waterlogging, bacterial mineralization of sulphur from organic matter and reduction of sulphate cause hydrogen sulphide to be formed. Sulphate is adsorbed onto clay colloids but it is not bound as strongly, and therefore it is more readily available to plants, than adsorbed phosphorus. In addition to its application as a fertilizer, appreciable sulphur may enter grassland systems from atmospheric fallout of sulphur dioxide near the sea or down-wind of industrial areas, e.g. in southern Scandinavia. This supplies sulphur to the grassland but causes serious long-term acidification because the sulphur dioxide is oxidized to sulphate in the soil. Topsoil pH (in 0.2 M KCl) has been reported to be as low as 2.8 and generally below 5.5 in a *Juncus* meadow near the sea in Sweden (Tyler, 1971).

5.1.2
Efficiency of utilization of fertilizer

The amount of fertilizer required to maintain soil nutrients at an unchanged level is equivalent to the total amount of nutrient removed in animal products (R in Fig. 5.1) plus any losses associated with the recycling of plant and animal material through the detritus chain back to the 'available' soil compartment divided by the fractional efficiency of fertilizer use, to account for losses of fertilizer which occur mostly by leaching and in the case of ammonium, volatilization. Expressed another way, in a non-steady state situation where the available pool size S is changing, the fertilizer which needs to be applied to meet pasture requirements is:

$$F = \{f(S \rightarrow P) + f(F \rightarrow P) + \Delta S\}/U' \qquad (5.3)$$

or

$$F = (U \pm S)/U \qquad (5.4)$$

where U' is the fractional effective utilization, U or $f(S \rightarrow P)$ plus $f(F \rightarrow P)$ is uptake by the pasture via the soil and by direct absorption of the fertilizer by the foliage respectively, and ΔS is the net change in available element in the soil. In the long term pasture production may tend towards a steady state in which:

$$F = U/U' - f(D \rightarrow S) \qquad (5.5)$$

where $f(D \rightarrow S)$ is the amount of element being recycled within the system, i.e. the rate of breakdown of plant and animal wastes.

In both short- and long-term situations the amount of fertilizer (F) is approximated by U/U', where U is the total amount of elements taken up by the grassland in a year. This total uptake by the grassland can be calculated from figures of herbage growth and nutrient concentrations – weighted concentrations of elements taking cognizance of the various plant parts and seasonal changes in concentrations. Ranges of mineral concentrations such as those given in Table 5.2 may be adequate for this calculation.

The fractional effective utilization of applied fertilizer, U, may range from zero to unity. There is a vast literature which shows that U depends on the element, the fertilizer type, the soil type, the climate, the method of fertilizer placement or incorporation, the herbage species and the age and stage of development. To give two examples: the fractional effective utilization of nitrogen when it is applied as inorganic fertilizer may range from 0.4 to 0.8 and commonly be above 0.6 on well-managed irrigated grasslands although it can be as low as 0.1–0.2 following the surface application of a slurry of animal waste due

Table 5.2. *Elemental composition (per cent of dry weight) of perennial ryegrass and white clover*

	Ryegrass	Clover
Phosphorus	0.26–0.42	0.26–0.42
Potassium	1.5 –2.5	2.0 –3.1
Calcium	0.4 –1.0	1.3 –2.1
Magnesium	0.09–0.25	0.18–0.25
Sulphur	0.13–0.75	0.24–0.36
Sodium	0.10–0.70	0.12–0.41
Chlorine	0.39–1.3	0.62–0.91
Silicon	0.6 –1.2	0.04–0.12

Source: Modified from the range of means cited by Whitehead (1966).

particularly to losses through volatilization; the fractional loss due to volatilization was 0.36 in one Irish study (Sherwood, 1983). For phosphorus the efficiency of fertilizer utilization is generally 0.7–1.0 (Karlovsky, 1983) although it may fall to almost zero when phosphorus application on silicious sands is followed by heavy rainfall. The fractional utilization of applied nutrients in a 'typical' temperate grassland grazed by dairy cows in New Zealand was calculated by Middleton & Smith (1978) to be approximately 0.8, 1.0, 0.7 and 0.75 for phosphorus, potassium, sulphur and magnesium respectively.

5.1.3
Amount of fertilizer required

In stable systems in which there is little likelihood of a change in element concentration in the soil, the total amount of compound fertilizer or manure which needs to be applied to meet the requirement for a particular element can be estimated simply by dividing F by the proportion of element within the fertilizer mixture. For example, if the amount of grass grown is 15 t per ha per year we could assess its requirement for potassium uptake as $15\,000 \times 0.02$ (from Table 5.2) or 300 kg K per ha per year. Then the amount of fertilizer required might be $0.3 \times 300/0.9$ assuming that the fertilizer programme has to replace 30 per cent of the total potassium, this percentage being removed by hay-making or lost during recycling through leaching, etc., and that the efficiency of application is 90%, i.e. 10% of applied fertilizer does not effectively enter into the nutrient cycle in the grassland. This amount of fertilizer (112 kg K per ha per year) would be provided from 300 kg potassium sulphate per ha or 80 t (fresh weight) swine manure per ha.

5.2
Uptake by plants

5.2.1
Uptake of inorganic elements

The amount of available element in the soil, the buffering capacity of the soil and the transport of the specific ion through the soil are external factors which affect the concentration of the ion at the root/soil interface and the speed with which the pasture takes up the element. Ions reach the root by three independent processes:

 (i) mass flow, through movement with water brought about by a gradient in water potential;
 (ii) diffusion, due to concentration gradients of the specific ions; and
 (iii) interception by the root or associated fungi (mycorrhiza) or bacteria, growing through the soil.

Mass flow can account for the total nitrogen, sulphur, calcium and magnesium required by most plants but cannot supply all minerals and does not supply all the nitrogen during periods of rapid growth. Transport of other minerals, e.g. phosphorus and potassium, is largely by diffusion (Clarkson, 1981).

Once an ion reaches the root, its rate of uptake is given algebraically by the product of three factors:
 (i) The concentration of the ion at the root surface, C_r.
 (ii) The surface area of the root: in practice this is the length of root (L) which may be measured per plant or per area of soil surface multiplied by twice the average radius of all the roots in the root system $\bar{a}$. The bar placed over the symbol indicates that we are referring to an average for the whole root system.
 (iii) The absorbing power, or efficiency of uptake, or apparent root transfer coefficient $\bar{\alpha}$.

That is (Milthorpe & Moorby, 1979):

$$U = 2\pi \times \bar{a} \times \bar{\alpha} \times L \times C_r \qquad (5.6)$$

A measure of the strength of root demand for an element is given by $\bar{a} \times \bar{\alpha}$. This demand varies between herbage species and with the age of the root system; it also varies appreciably between minerals for a particular species. Nonetheless the mathematical description does indicate that one species may have a relatively high 'root demand' for a particular element (e.g. *Stylosanthes* spp. for phosphorus: Probert, 1984) because of either $\bar{a}$ or $\bar{\alpha}$ or both. Further, the apparent root transfer coefficient can be related to the growth rate of the grassland (Nye & Tinker, 1969):

$$\bar{\alpha} = (W/W_r) \times (C_m/C_r) \times (R_w + R_c) \qquad (5.7)$$

where W is the weight of the whole plant, W_r is the weight of the root, C_m is the mean concentration of the element in the plant and R_w and R_c are the relative rates of change in the dry weight of the plant (the relative growth rate) and in the concentration of the element within the plant respectively. In contrast to $\bar{\alpha}$, C_r is only indirectly related to the growth rate and it is likely to be influenced more by soil properties and whether the element is highly mobile, e.g. nitrogen, or adsorbed onto soil particles, e.g. phosphorus.

Root morphology and the response to applied elements differ among herbage species. For example, subterranean clover (*Trifolium subterraneum*) roots are thicker and have less root hairs than those of ryegrass (*Lolium rigidum*) (Barrow, 1975). There is also variation within pasture groups, e.g. tropical legumes, in root morphology (Crush, 1974). It is currently thought that an unbranched root system is advantageous for water and mineral uptake during early growth whereas a fine, many-branched system is optimal in late growth (Fowkes & Landsberg, 1981).

The root system may be more efficient when it is infected with mycorrhiza (to facilitate phosphorus uptake), free-living rhizosphere bacteria such as *Azospirillium* (to facilitate nitrogen uptake) and bacterial symbionts, e.g. legume rhizobia. There may be a correlation between root structure (hairiness) and the ability of a pasture to accept and benefit from mycorrhizal infection: species with relatively few root hairs are most responsive (in terms of growth) to inoculation with mycorrhiza (Crush, 1974). Growth and responsiveness to applied inorganic fertilizer depend on whether roots are infected with mycorrhiza and bacteria, e.g. rhizobia. Clover may appear to be less responsive to applied phosphorus than is ryegrass but this lack of responsiveness is overcome by infecting the clover with mycorrhiza (Fig. 5.3).

5.2.2
Cation–anion balance

Plants may be placed into three groups with respect to the relative amounts of cations and anions they take up from the soil:

(i) Plants of the Chenopodiaceae (common rangeland species, e.g. saltbush) which absorb equivalent quantities of nutrient cations and anions so that their uptake of ions has no effect on the pH of the soil.

(ii) Grasses and ineffectively-nodulated legumes which on an equivalence basis absorb more anions than cations when nitrate is the main source of nitrogen, thereby causing an increase in the pH of the soil, and which absorb more cations than

anions when ammonium is the main source of nitrogen, thereby causing a decrease in the pH of the soil.

(iii) Nodulated legumes which acquire nitrogen by dinitrogen fixation and a few crop species, e.g. buckwheat (*Fagopyrum esculentum*) which take up nitrate nitrogen in such a way that uptake of cations exceeds that of anions, thereby causing a decrease in the pH of the soil.

Buckwheat belongs to the same plant family (Polygonaceae) as the weeds dock and sorrel and they too lead to a decreased soil pH. Differences between these three groups of plants have agronomic implications, most particularly the widespread acidification of soil under legume pastures.

Nitrate is usually the main source of nitrogen taken up by the plants in the first two groups (grasses and ineffectively-nodulated legumes). The uptake of nitrate is controlled metabolically (for a review see Hocking, Steer & Pearson, 1984) and accompanied by co-transport of cations, e.g. K^+, decarboxylation of acids within the root and extrusion of anions to the soil to maintain electrical neutrality within the plant. Proton (H^+) extrusion is relatively small so that the pasture itself causes little change in the pH of the root zone. However, addition to the soil of nitrogen fertilizers which contain ammonium and sulphate, and leaching of nitrate from the root zone, may cause long-term acidification, as mentioned in Section 5.1.1.

5.2.3
Fixation of nitrogen by legumes

In effectively-nodulated legumes, some nitrogen is supplied to the host plant through fixation of dinitrogen gas from the soil atmosphere into amino acids within the root nodule. The nutrition of legume–rhizobium symbiosis has been reviewed by Andrew & Kamprath (1978) and Pate & Atkins (1983). Some agronomic aspects are:

(i) There is a specificity between host and rhizobium (Table 5.3).

Fig. 5.3. Responsiveness to phosphate differs between legumes (a) (*Trifolium subterraneum*) and grasses (b) (*Lolium rigidum*). Lack of response to low concentrations of phosphate by legume can be overcome by inoculation with mycorrhiza. Closed symbols represent plants without mycorrhiza, open symbols represent inoculated plants; three levels of iron fertilizer were 0 (▲), 20 (■) and 40 (◆) g Fe(OH)₃ per pot. (From (a) Barrow, 1975; (b) Bolan, Robson & Barrow, 1983.)

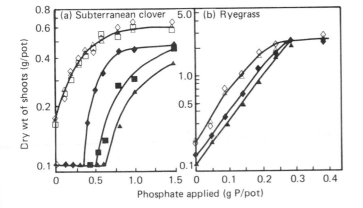

Table 5.3. *Host–rhizobium specificity*

Host	Rhizobium
Melilotus, Medicago	*R. meliloti*
Trifolium	*R. trifolii*
Pisum, Vicia	*R. leguminosarum*
Phaseolus	*R. phaseoli*
Glycine	*R. japonicum*
Lupinus	*R. lupinii*

Source: Mengel & Kirkby (1979).

(ii) Within an apparent single rhizobium species there are numerous isolates or lines which also show varying host specificity (e.g. Hardarson, Heichel, Vance & Barnes, 1981; Hardarson, Heichel & Barnes, 1982).

(iii) The colonization of the host and dinitrogen fixation by the rhizobia (collectively termed 'effective nodulation') depend on the soil type and the climate. Rhizobia are killed when placed in contact with fertilizer (e.g. Date, Batthyany & Jaureche, 1965). Nodulation is sensitive to soil pH, temperature and salinity.

(iv) The amount of nitrogen fixed by the rhizobia varies enormously according to species and growing conditions: values of 100–150 kg N per ha per year are common (Frederick, 1978) and up to 850 kg N per ha per year has been reported (Mengel & Kirkby, 1979).

(v) The proportion of its total nitrogen which the legume receives from the rhizobia appears to range from 0.3–0.6 in highly fertile situations (Heichel, Barnes & Vance, 1981) to virtually 1.0 in soils of low fertility.

(vi) The rhizobia and legume–rhizobium symbiosis have requirements for elements which differ quantitatively from those of the legume alone when growing on inorganic nitrogen. Dinitrogen fixation has a relatively high requirement for molybdenum, cobalt and calcium. The calcium requirement of the legumes themselves is also appreciably higher than that of grasses (Table 5.2). Conversely the rhizobia appear to have a relatively low requirement for phosphorus, sulphur and zinc (for a review see Robson, 1983).

(vii) Fixation of dinitrogen gas involves energy. The energy cost is difficult to estimate but it is probably about 5.1 mole carbon dioxide per mole ammonium: 0.85 moles of glucose are oxidized by the legume to produce energy for the production,

transport and utilization of each mole of rhizobial ammonium (Mahon, 1983). Others (reviewed by Phillips, 1980; and Gordon, Ryle, Mitchell & Powell, 1985) report that the cost of the entire dinitrogen-fixing process varies widely, from 0.3 to 20 g C per g N. Because this energy cost may be higher than that associated with the uptake and use of nitrate, small yield increases can be obtained by growing legumes on inorganic nitrogen. There is, of course, a high monetary cost involved in such substitution.

(viii) In terms of the total amount of energy used by the grassland system (rather than by only the herbage component) a grass–legume mixture is more efficient than a system based on grass plus fertilizer nitrogen (Table 5.4).

(ix) Application of nitrogen in inorganic fertilizers will increase the yield but decrease the proportion of nitrogen supplied by the rhizobia and possibly decrease the absolute amount of nitrogen obtained from the rhizobia (e.g. Mahon, 1983).

(x) Where legumes are largely acquiring their nitrogen by dinitrogen fixation, the uptake of cations exceeds that of anions by as much as five-fold, in contrast to a cation:anion balance close to unity for legumes growing on nitrate nitrogen (Israel & Jackson, 1978) and for grasses and non-legumes (e.g. Kirkby & Knight, 1977). Where the net uptake of cations exceeds that of anions there is no net extrusion of hydroxyl ions into the soil solution so that the soil becomes more acidic; it has been calculated that legumes may acidify their rhizosphere at a speed of 1 meq H^+ per g total plant dry weight increment (Israel & Jackson, 1978).

5.2.4
Soil acidification by legume pastures
Long-term acidification of the soil occurs under legumes. Decreases in soil pH of 0.5 units have been

Table 5.4. *Comparison of energy utilization between nitrogen-fertilized grass pastures and grass–legume pastures*

	Grass + Clover[a]	Grass + 450 kg N[b]
Total energy input (MJ)	6814	37940
Output (kg DM/ha)	9700	13200
ME output (MJ)	106700	145200
ME output/MJ input	15.7	3.8

[a] One application/year (kg): N, 60; P_2O_5; K_2O, 140.
[b] Five applications/year (kg): N, 450; P_2O_5, 90; K_2O, 190.
ME, metabolizable energy; DM, dry matter.
Source: Gordon (1980).

reported under white clover (*Trifolium repens*) within 4 years and a decline of 1 unit within 50 years appears to be widespread. Haynes (1983) lists the processes which contribute to acidification as (i) the accumulation of organic matter and a consequent increase in the cation exchange capacity and exchange acidity of the soil, (ii) the generation of mineral acids, e.g. nitric acid, during nitrification by soil microorganisms, (iii) the production of organic acids from carbon dioxide respired by soil fauna and flora (and to this we might add respiration from the roots and nodules), (iv) the absence of hydroxyl ion extrusion by legumes (see (x) above) and (v) the input of acidifying substances from the atmosphere.

Acidification of the soil changes the relative availability of elements in the soil. Aluminium and manganese become highly available (Fig. 5.4) and molybdenum becomes less available at pHs below 5.2. Both acidification and the altered balance of aluminium, manganese and magnesium thus indirectly decrease the effectiveness of rhizobia. This leads to the sequence:

long-term legume and/or high use of nitrogen fertilization → soil acidification → aluminium or manganese toxicity → decline in herbage growth and difficulty in re-establishing pasture.

There is therefore need for concern and research into the sustainability of legume-based systems. Current agronomic recommendations for amelioration of soil acidification include deep ripping, application of lime and the establishment of grasses tolerant to acid soils. However, there are differences among legumes in their ability to grow at low pH or in relation to changes in availability of aluminium and calcium. This has led to the recognition that some legumes can be

recommended as more tolerant to acidity than others. For example, Munns (1978) cites some tropical legumes, e.g. *Centrosema* and *Macroptilium* as highly tolerant, other tropical legumes, e.g. *Desmodium* and *Phaseolus*, and the temperate *Trifolium subterraneum* as moderately tolerant, and *Medicago*, *Pisum sativum* and *Trifolium repens* as highly sensitive to acidity. Future management of soils made acid by prolonged use of legumes, nitrogen fertilization or industrial fallout will presumably include the adoption of appropriate legumes.

5.3
Distribution within the plant

Elements may be classified as mobile, variably-mobile or immobile within the plant: some elements are metabolized and translocated to new growth when the plant is young or in late stages of senescence, but they seem to be immobile during most of the life of the grassland. Further experimentation is required to define these classes; at present they are (J. Loneragan & A. Robson, personal communication):
(i) Mobile: nitrogen, phosphorus, potassium.
(ii) Variably-mobile: copper, zinc, sulphur, magnesium.
(iii) Immobile: calcium, boron, manganese, molybdenum.

Mobile elements, e.g. nitrogen, do not move in one direction but cycle according to the sites of metabolic activity in the plant. The cycling is most apparent in grasses and determinate forage legumes, e.g. lupin (*Lupinus*), where vertical gradients of element concentration occur due to a vertical canopy structure and the determinate habit of the pasture. Flow diagrams of nitrogen distribution in wheat and lupin are given by Simpson, Lambers & Dalling (1983) and Pate & Atkins (1983) respectively.

Immobile elements move into growing tissues and stay there. That is, the concentration of the element in new growth depends on the current rate of uptake and immobile elements are lost from the pasture when the organ dies and falls off.

Fig. 5.4. Trends in characteristics of a granitic soil with time under subterranean clover pasture. (a) pH; (b) manganese extracted in 0.01 M $CaCl_2$ and (c) exchangeable aluminium. V, virgin soil; C and D, soil after 26 years under pasture; O, soil after 55 years under pasture. (From Bromfield *et al.*, 1983.)

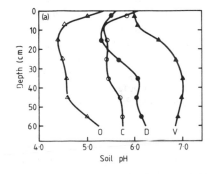

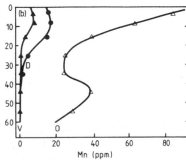

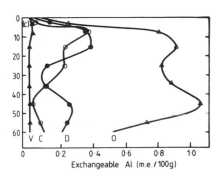

The mobility of an element has consequences for both the diagnosis of element sufficiency in grasslands and for the management of fertilizer application and grazing. For diagnosis of element status using plant samples, a whole-plant sample is most likely to estimate accurately the status of mobile elements since these are likely to be found at reasonable concentrations in new growth irrespective of current uptake or soil status; a sample of new growth will estimate best the current status of immobile elements. Management should adopt fertilizer practices which ensure a continuous supply of immobile elements whereas the time when mobile elements are available is less critical for sustained grassland growth. Conversely, some mobile elements, particularly nitrogen, may be accumulated at high concentrations so that to avoid animal disorders grazing should not follow shortly after application of nitrogen fertilizers.

The relative concentrations of an element in different plant parts changes with developmental stage and in response to management, e.g. defoliation. The change in the concentrations of each element is discrete and not necessarily related to that for other elements. In uncut plants of effectively-nodulated phasey bean (*Macroptilium lathyroides*) nitrogen accumulates mainly in the leaves and stems prior to and during early flowering; pods are the major sites of nitrogen (and carbon) accumulation during pod filling when roots and nodules gain only 10 and 20 per cent of the total nitrogen and carbon fixed by the plants (W. Othman, personal communication). Defoliated plants which retain developing pods suppress regrowth through the mechanism of effective competition by the pods with new shoots for nitrogen and with nodules for carbon. Ageing leads to senescence, at which point the element re-enters the greater cycling of elements by passing to the dead material compartment.

5.4
Senescence and element release from dead material

The detritus chain involves the movement of material through the dead material compartment, as in Fig. 5.1. The dead material is functionally four pools: standing dead plant material, unincorporated (plant) litter, incorporated litter and animal faeces.

5.4.1
Herbage death

The standing dead plant pool is important agronomically because it affects herbage growth and animal intake (Chapter 6). This pool may fluctuate in size according to seasonality of the weather, with greater senescence during periods of extreme temperatures

and drought but some leaf and root material entering the dead pool throughout the year. The size of the pool also follows closely the life of the pasture plant; a peak of senescence is particularly obvious in annual species. Nevertheless, few experiments have quantified herbage death, $f(P \rightarrow D)$, in such a way as to allow generalizations, despite Hunt (1965) and others drawing attention to its quantitative importance. In one study, of a grassland consisting of perennial ryegrass, white clover and *Poa annua*, the rate of leaf lamina senescence per unit area (S') increased linearly with herbage mass (W_s) (Bircham & Hodgson, 1983):

$$S' = 0.5 + 0.023W_s \qquad r^2 = 0.91 \qquad (5.8)$$

where S' is in kg per ha per day and W_s is in kg per ha.

Rates of plant death and litter fall are most sensitive to temperature and increase during periods of low available soil water. Death rates, $f(P \rightarrow D)$, of 0.1 per cent and 7.5 per cent per week in winter and summer respectively are considered appropriate for phalaris–white clover grasslands in Australia. The higher value is approximately equivalent to 40 kg per ha per day, which is reported for ryegrass in England. Relative rates of disappearance of leaf litter of 4 mg per g litter per day have been reported in a range of grassland situations (e.g. Yates, 1982).

5.4.2
Dung and urine

Dung deposition is related to the number and type of grazing animal; it is non-seasonal. It declines per animal with the decline in herbage intake as the stocking rate is increased: the wet weight of dung decreased at the rate of 0.16 kg per cow per day per unit increase in stocking rate to 8.6 cows per ha in Victoria (Stockdale & King, 1983). Typical rates of deposition of elements in dung pats are equivalent to 650–850 kg N, 125–400 kg P and 150–170 kg K per ha per year.

In grasslands grazed by cattle, the area rejected due to fouling by dung and urine may reach as much as 45 per cent of the total area (Marsh & Campling, 1970). Elements, especially potassium, become concentrated on animal camp sites. This results in major changes in botanical composition (Section 7.1). Such redistribution of elements was found to have a greater effect on productivity than long-term (1948–76) application of different rates of superphosphate. Redistribution frequently results in potassium deficiency in non-camp areas of paddocks.

5.4.3
Decomposition

Once material enters the dead pool it is decomposed by organisms as reviewed by Bell (1974) and

Lodha (1974). The organisms are surface-dwelling bacteria, the majority of which can be found on the plant before death: yeasts; filamentous fungi, which likewise can be isolated from live, particularly old, plant parts; and internal colonizers such as saprophytic and parasitic fungi. The fungi which decompose dung, or their spores, may be present on the food when it is eaten and certainly as it is defecated. They germinate when the dung is deposited but, although growing together in the decomposing faeces, the various species of fungi have differing lengths of life cycle so that the fruiting bodies of various fungi appear as a sequence during decomposition of the dung. This sequencing of visible presence, together with colonization and recolonization, gave rise to the term 'succession' to describe the changes in decomposer populations during decomposition of organic material.

Soil fauna such as arthropods and earthworms accelerate decomposition but they are not essential for it to occur. Fauna also have a role in 'stirring' the litter and the soil. This aids litter decomposition, soil aeration and structural amelioration. Grasslands have faunas which are characteristic of the plant species, climate and microclimate, and management. Seasonality of weather has a dominant effect on the composition of the invertebrate population and there are also short-term changes in abundance associated with grazing, cutting or conservation: collembola and acarine numbers fall when the sward height is reduced by heavy grazing or cutting whereas abundance, particularly of larger insects, increases during forage conservation (Purvis & Curry, 1981).

As Springett (1983) has shown, earthworm species vary in their rate and mode of activity such that some are more effective than others in the mixing of soil and nutrients. *Lumbricus terrestris*, the most common earthworm, selects litter with the highest nitrogen and soluble carbohydrate content and draws it to

the soil surface prior to eating it. Curry & Bolger (1984) found that the presence of *Lumbricus terrestris* in reclaimed peat increased the rate of decomposition of litter placed in cages from 5 per cent per month in their absence to 15 per cent per month in their presence.

Decomposition is most simply described by a skewed bell-shaped curve in which the rate of decomposition declines rapidly with time after death or deposition. The rate of decomposition varies according to element, type (e.g. leaf versus stem) and species of material; environment, particularly temperature and available soil water; and microbial species. Ross, Henzell & Ross (1972) used a mathematical expression for the rate of decomposition to predict a sequence of leaf death and decay of grass and legume where a first peak of legume death was due to shading by the grass and was followed by a peak of grass death due to the grass running out of nitrogen and later by a rise in legume senescence. Arising from this death sequence, the transfer of elements from legume litter to the available soil pool $f(D \rightarrow S)$ occurred gradually: appreciable transfer (>10 kg/ha) took at least 100 days.

Recently Vallis (1983) used ^{15}N techniques to follow decomposition of legume litter ($f(D \rightarrow S)$) and uptake of released nitrogen by companion grass ($f(S \rightarrow P)$). This study found that 25–91 per cent of ^{15}N in dead legume was lost from the organic material pool in one year but only 7–25 per cent was recovered in the companion Rhodes grass. There was a curvilinear relationship between nitrogen uptake by grass and the nitrogen concentration of the applied legume litter (Fig. 5.5). This illustrates the importance of microflora and their environment on $f(D \rightarrow S)$. Microbial activity is limited where litter nitrogen is low whereas high nitrogen concentrations stimulate microbial activity and cause an excess of available nitrogen above the needs of the detritus organisms.

During decomposition, the dry weights of litter and its components decline although the protein and element contents of litter (or faeces) may increase during decomposition due to the growth of microorganisms in and on the litter. It is estimated that microorganisms use 1 kg N from the soil or added

Fig. 5.5. Cycling of legume nitrogen. (a) Loss after 1 year of ^{15}N from legume litter placed on the soil surface and (b) its recovery in Rhodes grass, in relation to nitrogen concentration in the litter. The litter consisted of the stems and leaves of two species. (From Vallis, 1983.)

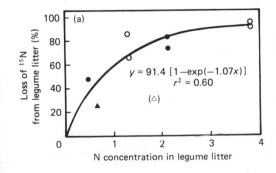

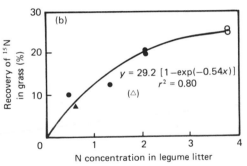

residue to convert 100 kg of residue to humus. A decline in soluble organic material such as amino acids in litter is usually not apparent until after some time, perhaps one month, due to the flow of material through the soluble pools. The dry weight of material which is relatively indigestible, e.g. lignin, declines but its percentage contribution to the remaining dry weight increases during organic matter decomposition.

This leads to mention of the efficiency of the detritus organisms in releasing elements and the retention of these elements in the available soil pool. Sources of losses of elements from the grassland system are shown schematically in Fig. 5.1. The detritus chain does not, in itself, confer inefficiencies or loss of elements although it is often the largest compartment of elements in the system and in this sense renders them 'unavailable' for herbage growth. Dung beetles, native to South Africa, accelerate dung decomposition and thus reduce the size of the dead material pool; however, beetles do not necessarily decrease the losses of elements as they pass through the detritus chain. The losses of elements as they pass through the detritus chain are due to volatilization and leaching and vary between elements. Herbage budgets, as a rule-of-thumb, assume that 20–25 per cent of recycled nitrogen will be lost in these ways whereas virtually no sulphur is lost, i.e. losses of sulphur are small and replenished from sources within the system. The extent of loss and the speed with which the element becomes available (enters compartment S in Fig. 5.1) depend on the form of the dead material. For example, as much as 94 per cent of sulphur from urine is quickly available for plant uptake compared with about 50 per cent of sulphur from faeces and virtually none of the sulphur in plant litter (Kennedy & Till, 1981).

5.5
Animal intake and product removal

The intake of various elements by grazing animals varies between elements and between animals and according to the amount of available feed and to plant morphology and quality (Chapter 6). Table 5.5 shows intake for a lactating cow and the fate of elements in milk and excreta. Likewise, the loss of elements through product removal, the loss from the system of elements by removing milk, beef, hide, wool and dung, depends on the concentrations of elements within meat, wool and milk and these vary between elements and among animals (Table 5.6). For the purpose of budgets concerning mineral nutrition and livestock management it is appropriate here to note only that these mineral concentrations, when multiplied by animal liveweights, permit us to calculate fluxes of minerals within the grassland system and the amount of fertilizer which may be required to maintain these fluxes.

5.6
Losses from the system

Losses of elements by leaching, run-off and erosion and, in the case of nitrogen, volatilization, are accelerated by grazing animals. This is because animals produce very high local concentrations of elements in dung and urine, e.g. ammonium nitrogen in the surface soil can approach 1000 ppm during the first day after urine application (Vallis et al., 1982). Recovery of nitrogen in the soil following application of urine ranges widely but is often about 60–70 per cent (Ball & Keeney, 1983). Moreover, grazing and treading increase run-off and erosion. Leaching, run-off and erosion losses of phosphorus from fertilized grassland usually range from 2 to 10 kg per ha per year: losses by

Table 5.5. *Mineral composition of herbage intake by lactating cows consuming 16 kg of dry matter, the daily excretion of minerals in milk, faeces and urine and the percentage of mineral intake not excreted in the faeces (apparent availability)*

Element	% of herbage DM	Daily intake (g)	Daily excretion (g) Milk[a]	Faeces	Urine	Availability (%) Mean	Range
Potassium	3.02	483	41	53	389	89	80–95
Sodium	0.37	59	10	9	40	85	66–92
Chlorine	1.08	173	29	21	123	88	71–95
Sulphur	0.42	67	7	18	42	73	64–82
Calcium	0.61	98	30	68	1	30	16–47
Magnesium	0.23	37	3	30	3	17	7–33
Phosphorus	0.41	66	24	48	0.2	27	10–46

[a] Refers to 25 kg milk per day.
Source: Kemp & Geurink (1978).

these pathways may amount to 8, 10 and 18 kg K, Mg and S per ha per year respectively (Middleton & Smith, 1978).

Leaching of nitrogen may amount to up to 100 kg N per ha per year in highly-productive grazed legume under relatively high rainfall (Simpson & Steele, 1983). However, leaching of nitrate from soil under grassland is often low (about 3 per cent of the nitrogen taken up by the herbage) whereas leaching under annual forages may amount to 10–30 per cent of the nitrogen taken up annually by the crop (Jaakkola, 1984). This difference is due presumably to the grassland having a longer season of uptake of nitrogen and a greater ability (owing to high root and soil organic matter) to trap ions in the surface soil. Leaching of ions away from roots during a season has been described as depending on root depth and soil characteristics:

$$L \sim \frac{[NO_3] \times D \times P \times 0.5d}{100 \times (P + \theta)} \qquad (5.9)$$

where D is the effective rooting depth (cm), $[NO_3]$ is the residual nitrate nitrogen in the top metre of the soil (kg per ha) at the beginning of the season, P is the accumulated drainage (cm, the total rain or irrigation water moving through the soil) and θ is the volumetric field capacity of the soil (cm per cm) (Burns, 1980). Losses of nitrate due to leaching are 2–4 times greater and sometimes up to 10 times greater under legume pastures than under grasses (Haynes, 1983). This is attributed to greater nitrification and less immobilization of nitrogen by microorganisms in high-nitrogen legume material than in grass litter.

Volatilization can be an important pathway of loss of nitrogen following application of ammonium fertilizer. Ammonia loss occurs directly from plants, fertilizer, the soil, plant litter and excreta. Direct loss from the grassland depends on the concentration of ammonia in the atmosphere and the age and species of the plant. Direct losses are likely to be small (Section 5.1.1). Denmead, Freney & Simpson (1976) found that most of the ammonia released from the soil was absorbed by the pasture. By contrast, volatilization from spread fertilizer can be appreciable and rapid in hot climates: Catchpoole, Harper & Myers (1983) found that 20 per cent of applied urea nitrogen was lost within 14 days of fertilizer application to subtropical grasslands, and that these rapid losses could be as high as 42 per cent. These values compare with a total nitrogen loss of 50 per cent by all pathways.

5.7
Element deficiency and fertilizer needs

There is a basic relationship between the content of elements in a grassland and its growth or yield. This is shown in idealized form in Fig. 5.6. When the element content (the total amount of an element in the plant sample) or concentration (the amount expressed as a percentage or fraction of the dry weight) is very low, either absolutely or in relation to other elements, then growth is low. Increased availability of elements or the overcoming of an imbalance between elements causes growth to increase and this may bring about a slight decrease in concentration due to the dilution of elements in new growth. In the next stage growth increases while the element concentration remains unchanged, i.e. growth is proportional to the availability of the element. Then as the availability increases so do the growth rate and the element concentration in the herbage until the so-called critical level is reached. Thereafter, growth does not change but the element content increases, e.g. nitrate accumulates perhaps to levels which may cause animal disorders (Chapter 6). High levels of elements impair growth (Fig. 5.6a).

Element deficiencies may be defined at clinical, sub-clinical and economic levels; element deficiencies or imbalances may be diagnosed, at least at the clinical level, by:

(i) Prediction based on budgeting from the history of fertilizer use and the efficiency of utilization of the fertilizer and thus its residual value. Since this is an aspect of management it will be dealt with in Section 5.8.

Table 5.6. *Minerals in animals and animal products (per cent of liveweight)*

	Calf	Steer	Lamb	Sheep	Unwashed wool	Cow's milk
Nitrogen	2.6	2.4	2.5	2.5	11.4	0.61
Phosphorus	0.67	0.68	0.49	0.45	0.03	0.10
Potassium	0.17	0.15	0.15	0.12	4.7	0.12
Sulphur	0.15	0.15	—	0.15	3.5	0.04
Calcium	0.12	1.3	0.91	0.84	0.13	0.11
Magnesium	0.15	0.03	0.03	0.03	0.02	0.001

Source: Various authors, mostly collated in Wilkinson & Lowrey (1973).

(ii) Symptoms seen in the plant shoot. By the time symptoms are obvious and unambiguous (and they may never be the latter) at least 30 per cent of the growth rate may have been lost due to element insufficiency or imbalance (A. D., Robson, 1984, personal communication). Nonetheless guides to symptoms are available for some pasture grasses and legumes in the Australian CSIRO Technical Papers series (e.g. Smith & Verschoyle, 1973). Diagnostic keys to element deficiencies in subterranean clover (Snowball & Robson, 1983) indicate the scope and complexity of using visual symptoms to diagnose the element status of the plants. Some deficiencies (e.g. of molybdenum) or toxicities (e.g. of aluminium or manganese) may be seen as failure of effective nodulation and stunting of legume roots.

(iii) Soil analyses. These are usually specific to combinations of soil type and climate and they do not always correlate with plant concentration or performance. Nonetheless they are widely, and usefully, employed and will be discussed later in this section.

(iv) Biochemical assays. We are aware of only one such assay, i.e. the use of ascorbic acid oxidase as an assay for copper deficiency in subterranean clover. Biochemical assays are presently not robust but they have some potential in being rapid and perhaps unequivocal, i.e. element-specific.

(v) Plant analysis. Concentrations of element within the plant are more directly related to herbage growth than are concentrations within the soil. Nonetheless, as mentioned earlier (Section 5.3) and detailed in Mengel & Kirkby (1979), account should be taken of plant age, the plant part which is most appropriate to sample and the critical concentration above which there may be no growth response to further additions of fertilizer. The critical concentration varies between pasture species: common values for legumes are 0.16–0.24 per cent P, 0.7–0.8 per cent K and 0.15–0.18 per cent S (Mott, 1978). Unfortunately the critical concentrations change (decline) markedly with plant age and they may also be affected by climate and soil type to an extent which we do not currently understand. Moreover, it is not clear whether the critical concentrations for growth coincide with those for animal health.

The relationship between element content in the pasture (i.e. concentration × biomass) and availability in the soil is generally asymptotic (Fig. 5.6b). Thus plant and soil analysis are equally sensitive at low levels of soil elements whereas soil analysis may be most appropriate in the higher range of element availability.

5.8
Implications for grassland growth and management

Fertilizer, grassland and livestock management are carried out for a purpose, which may be to ensure plant establishment, increase or maintain grassland growth, achieve a particular level of herbage quality or alter the botanical composition. Any or all of these might, of course, be carried out in the expectation that they will lead to an increase in animal production. Here we will place emphasis on fertilizer management and grassland growth; longer-term effects of nutrition, particularly on botanical composition, will be considered briefly in Section 7.1.

5.8.1
Amount of fertilizer

The amount of fertilizer needed for either the most economic or the maximum grassland growth

Fig. 5.6. (a) Idealized relationship between herbage growth and concentration of elements in the pasture. (b) Idealized relationship between element content of herbage and element concentration in the soil solution. (From Mengel & Kirkby, 1979.)

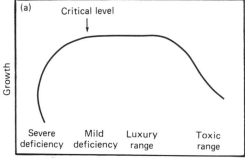

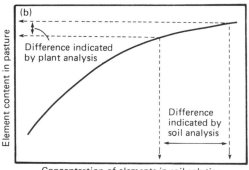

may be viewed positively, as the amount of fertilizer which should be added to achieve a particular rate of growth, or negatively, as the extent to which growth is constrained by the current availability of elements. From either viewpoint, fertilizer requirements are considered on the basis of (i) annual grassland growth, (ii) seasonal growth or (iii) growth over a short term, e.g. the interval between grazing.

Analyses of the extent to which nutrition constrains annual grassland production may involve surveys. These equate maximum grassland production with that currently being achieved on the best class of grazing land within a district or in a comparable but perhaps more highly developed district. The survey method makes no pretention of predicting potential productivity but, usefully, relates current livestock-carrying capacity or grassland growth to the maximum found in the survey area. This approach is empirical and it is of little help in diagnosing which nutrients are limiting grassland growth, predicting marginal returns to application of fertilizer or, indeed, assessing whether the limitation to growth might be due to soil factors other than nutrition.

Quantification of the extent to which growth is constrained by nutrition may be more mechanistically made by defining a nutrition index in analogous terms to a temperature index (Fig. 3.10). The nutrition index (NI) ranges from zero, when there is no growth, to 1, when growth is not restricted by nutrition. The nutrition index may be related to the availability of a particular element in the soil at some particular site. The relationship is asymptotic:

$$NI = 1 - b \exp(-CX) \qquad (5.10)$$

where C is the coefficient of curvature of the response curve, b is an empirical constant to describe the responsiveness of the site, ranging from 0 to 1, and X is the amount of available element, which may be varied by applying fertilizer. Appropriate units are kg element per ha. Nutritional indices derived from experiments with phosphate and potassium fertilizer in Australia adhere to this general equation (Fig. 5.7). The coefficient of curvature is large when any given NI is reached at relatively low levels of available element: C was higher in experiments carried out in Victoria than in experiments performed in New South Wales, Australia (Fig. 5.7a). If C is considered in terms of plant response to applied fertilizer rather than to available element in the soil, then soils with a high buffering capacity due to sorption or fixing of the element have low C values. Sesquioxide sands with a high fixing capacity have C values for phosphate as low as 0.03 ha per kg of applied P whereas siliceous sands may have values as high as 0.2 ha per kg P (W. Bowden & D.

Fig. 5.7. Nutrition index (NI). Herbage growth expressed as a fraction of that under optimum nutrition for (a) phosphorus and (b) potassium. The phosphorus index is based on phosphorus fertilizer experiments in Victoria and New South Wales, Australia. (——), Total pasture yield; (---), legume yield. Extractable potassium refers to that extracted in 1:100 soil:0.5 N sodium bicarbonate solution from western Australia. (From: (a) I. Malling, 1983, personal communication, and D. B. Batten, 1977, personal communication; (b) W. J. Cox, 1977, personal communication.)

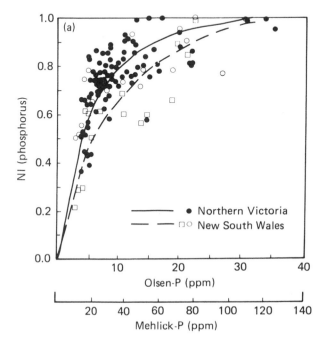

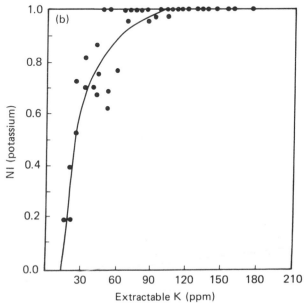

Bennett, 1974, personal communication). Incorporation of fertilizer into the soil gives a higher apparent C value than does topdressing, and the type of fertilizer applied may cause a five-fold variation in apparent C. The responsiveness of the grassland to available element also depends on the species, lupins being reported as having three times the phosphorus requirement ($C = 0.4$) of grasses such as wheat ($C = 0.12$) on virgin sands. For wheat, K. M. Curtis & K. G. Helyar (1983, personal communication) fitted C to soil phosphate buffer capacity (P) and average annual rainfall (R, mm):

$$C = 0.157 - 0.004\,97P - 0.000\,043\,5R$$
$$+ 0.000\,213R^2 \qquad (5.11)$$

The Nutritional Index, as defined above, relies on experimental data of crop or pasture yield and soil parameters. Such data exist from fertilizer experiments with single elements and these have led to models for predicting economic rates of phosphorus application for pasture growth (Wheeler, Pearson & Robards, 1987).

Single-element models for predicting strategies for the most economic application of phosphorus employ asymptotic curves to describe response to applied phosphorus and they incorporate the notion that soil phosphorus is partitioned between soil solution, rapidly-exchangeable and unavailable pools. The size of these pools, and plant responsiveness, depend on the amount of phosphorus which has been applied previously to the grassland. Plant responsiveness also depends on the soil type and on the efficiency of fertilizer utilization (Section 5.1.2). Models of responsiveness to nitrogen (e.g. Greenwood *et al.*, 1971) use an overturning curve to mimic toxicity at high concentrations. Because of the mobility of nitrogen, however, efficiency values (B values in Greenwood's terminology) are very site-specific.

As models for predicting fertilizer requirements become more mechanistic, it is important that they estimate the depth of the roots. This depth determines the total volume of soil from which the grassland may extract elements; it also delineates the lower level below which leaching, e.g. of nitrate, causes elements to be lost from the system. Rooting depth can be estimated only empirically (Burns, 1980; Pearson & Jacobs, 1985).

The prediction of yield responses to fertilizer becomes more complicated when more than one element is considered at a time. Greenwood's work with vegetables approaches a useful, generalized multi-element model (Greenwood *et al.*, 1971, 1980; Greenwood, 1981).

In intensive grassland systems, it is desirable to predict accurately the optimum rate of fertilizer application that will produce herbage between cuttings or grazings in the most economic way. Here, the available concentrations of elements in the soil are high and the primary aim is to maintain not only maximum growth but also high concentrations of element in plant tissues to maximize animal production. Element yields can be calculated from target concentrations and forage yields; these are the guide used for deciding the amount of fertilizer to be applied.

5.8.2
Timing of fertilizing

The most convenient and least costly method of fertilizing is a single annual application of fertilizer. This might be applied at the start of or during a period of rapid growth, e.g. in spring following grazing and accompanied by irrigation. This timing will ensure that utilization (U, Section 5.1.2) will be high. However, it will produce more feed at a time of year when there is probably an excess of feed over animal requirements. Effective management, then, is a cost-efficient compromise between single and multiple applications and between maximizing U and producing feed when it is actually needed: maximizing the conversion of applied element, e.g. nitrogen, into dry matter consumed by the grazing animal. Environmental factors, e.g. high temperatures, which would decrease U by excessive volatilization of ammonium, are usually used as criteria for decisions over a shorter time scale: the farmer will take account of the weather when deciding to apply fertilizer today or leave it until tomorrow.

The timing of application of fertilizer with respect to short-term forage production should take account of a lag period before there is a measurable stimulation of growth and the diminishing effect of fertilizer with time after its application. Responsiveness varies between species: leaf extension responds to nitrogen more in ryegrass than in fescue (Wilman & Mohamed, 1980). The lag period before nitrogen gives a measurable stimulation of growth is 2–21 days under adequate rainfall or irrigation and moderate temperatures; Read (1976) found the lag period to be 15 days for irrigated *Phalaris aquatica* on clay soil. Thereafter the declining effectiveness or residual value of the fertilizer is described by the slope of a negative hyperbola which, of course, reaches zero when the fertilizer has no residual value.

At the other extreme of extensive crop–pastoral rotations, farmers may choose to apply fertilizer only before the cropping phase. The cost of this fertilizer is usually recovered in extra crop yield and the residual, although diminishing, value of the fertilizer is available for re-establishing the pasture phase. In turn, the pasture phase, if dominated by a legume, increases soil nitrogen and organic matter, which improve crop yields. Lucerne, lupin and red clover have the reputation of producing high residual nitrogen in relation to their duration.

5.8.3
Type and method of application of fertilizer

Factors which affect these managerial decisions are the cost of fertilizer per unit of element, its ease of application, likely losses, e.g. through volatilization, its availability in soil solution for uptake by plants, the speed of plant response, and any long-term effects the fertilizer may have, e.g. on soil acidity. Fertilizers such as anhydrous ammonia which carry a single element, require special equipment for application, are relatively sensitive to environment (U depending on temperature, moisture, soil organic matter and calcium, and depth of application) and have undesirable long-term effects on the soil (acidification), are becoming less popular as farmers tend towards the use of stable, easily-applied compound fertilizers.

5.8.4
Grassland maintenance and restoration

On light-textured soils, continuous grassland may lead to desirable increases in soil organic matter and particle aggregation (Fig. 8.5). However, on soils which are high in clay or organic matter, intensive production and application of fertilizers over a long period (such as more than 80 years) can cause acidification and deterioration of soil structure (e.g. increased bulk density), accumulation of organic matter and a decrease in microbial activity (Shiel & Rimmer, 1984). There is thus a general trend towards increased organic matter and thus nutrient stability under grassland but these stabilizing effects may be offset by particle aggregation, acidification and uneven distribution of minerals caused by grazing giving rise to undesirable long-term effects on plant composition and productivity. Among these undesirable trends are invasion by acid-tolerant weedy taxa, dominance by unpalatable nitrophilous weeds near stock camps and inability to re-establish legumes. The undesirable trends lead to greater seasonality in grassland production, lower production and a lower proportion of production passing to the grazing animal.

Management can control the rate of occurrence of changes in species and soil characteristics. Changing the type or rate of application of fertilizer or placement of fertilizer at depth in environments where the topsoil may dry periodically, can contribute to maintenance of productivity (e.g. Kemp, 1984). Deep ripping (tyned ploughing to depths of 30 cm or more) may increase the availability of elements in the root zone and increase the rooting depth by creating vertical channels of soil with low bulk density.

The species composition can be altered by sowing, particularly by drilling directly into the existing sward. This has implications for nutrition, directly through fertilizing at sowing and, in the long term, through a fertilizer programme appropriate for the newly-sown species and in indirect ways such as the effects of sowing or oversowing on the proportion of legume in the pasture and on soil organic matter and structure. Direct drilling has associated with it an increased use of herbicides and possibly pesticides. These in turn affect the soil fauna and flora although we currently have little knowledge of the quantitative effects on fauna or their significance in nutrient cycling. In one study, Anderson, Armstrong & Smith (1981) found that application of fungicides, at 5 μg per g soil, caused transient decreases of 40 per cent in soil flora and a shift from fungi to bacteria while higher rates of fungicide application caused long-term decreases in microorganisms.

The coating of an inoculum of rhizobium in a lime pellet and, if the pH is below 5, banding of the inoculated, pelleted seed in lime, enhance the survival of sown rhizobium and increase the likelihood of legume establishment. The selection of inoculants which are tolerant to acidity and pelleting them with rock phosphate is recommended for species sown in the acid llanos of South America (CIAT, 1980).

The application of lime at rates in excess of 1 t per ha and usually 3–6 t per ha over three years has two effects on restoration. Oxide, carbonate and hydroxyl ions of calcium increase the soil pH. This in turn makes some elements more available for rhizobial dinitrogen fixation, e.g. molybdenum, and others less available, e.g. zinc and manganese. Calcium improves the soil structure through its adsorption (and displacement of Na^+ in saline soils) causing flocculation of the soil colloids. Clay soils are high in exchangeable H^+ and also most liable to compaction; these are most frequently in need of liming and, because of their high H^+ content, require more lime than coarse-textured soils for a given increase in pH.

Finally, adoption of improved, tolerant grassland species will play a role in grassland maintenance in nutritionally-marginal situations or renovation after degradation of soil fertility. For example, in tropical South America, the high cost of phosphorus fertilizers and high soil aluminium and acidity result in low levels of animal production. It has been shown recently that species differ in their tolerance to high aluminium and low phosphorus, probably through sensitivity of root elongation: *Stylosanthes guianensis* and *S. capitata* are tolerant of these nutritionally-marginal conditions whereas *Centrosema* and *Macroptilium* are highly sensitive (CIAT, 1980).

5.9
Further reading

Kemp, A. & Geurink, J. H. (1978). Grassland farming and minerals in cattle. *Netherlands Journal of Agricultural Science*, **26**, 161–9.

Mengel, K. & Kirkby, E. A. (1979). *Principles of Plant Nutrition*, 2nd edn. Berne: International Potash Institute.

6

Herbage quality and animal intake

This chapter deals with four factors which determine the amount of feed eaten by a grazing animal; three of these closely involve agronomy. The factors are quality, sward structure, availability (the amount of feed relative to the need of the animal) and the type, size and productivity of the animal (Fig. 6.1).

These factors should be considered together: they all operate within a given grazed field. Thus Stockdale (1985) related the daily intake of dry matter (DM) to six variables which are expressions of quality, composition, availability and animal type:

$$I_A = -7 + 0.27W_A - 0.0018(W_A)^2 + 1.1W$$
$$+ 6.2D - 0.63P_T + 0.11LW - 1.46DE$$
$$(r^2 = 0.89) \quad (6.1)$$

The numbers in Eqn (6.1) were found by curve-fitting and, as they are probably site-specific, need not concern us. More generally, the equation illustrates that the amount of feed eaten by a milking cow (I_A, kg per cow per day) can be predicted mathematically from the amount of available herbage per cow (W_A, kg DM per cow per day), the total amount of herbage (W, kg DM per ha, ranging between 2800 and 5200 kg per ha), the digestibility (D, per cent), the plant type (P_T, temperate or subtropical), animal liveweight (LW, kg, ranging between 320 and 420 kg per cow) and the duration of grazing (DE, days, ranging from 10 to 365).

Biologically, the apparent significance of variables such as those in Eqn (6.1) can be explained by underlying changes in herbage quality or, more specifically, by changes in the quality of the feed which the animal has been permitted to select or forced to eat. 'Quality' is a term which encompasses the chemistry and structure of the feed: it is a consequence of chemistry and anatomy.

6.1
The basis of chemical composition

Chemical composition, particularly as it relates to grazing animals, is often assessed by apparent digestibility (D):

$$D = 100 \times (I_A - f)/I_A \quad (6.2)$$

where I_A is the amount of feed consumed and f is the amount of faecal output. This is commonly calculated on all dry matter eaten by the animal (in which case it is called dry matter digestibility D or DMD). Equally, digestibility may refer only to organic matter (OMD), all minerals in the feed (for historical reasons called ash, hence ash digestibility) or to any specific component of the feed, e.g. cell wall digestibility (CWD).

Other measures of the chemical composition of feed, which are used in animal nutrition to describe

Fig. 6.1. Factors which affect animal intake. Rate-governing variables are shown as ▢◁; factors which contribute to the rate are given in boxes.

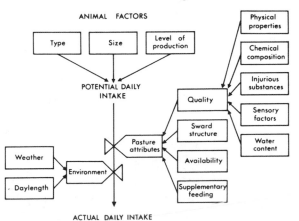

protein concentrations, are crude protein, and biological value. Crude protein (CP) is

$$CP = 6.25 \times \text{per cent N in feed} \qquad (6.3)$$

This simplification is based on two unsound assumptions: (i) that all nitrogen in the herbage is present as protein and (ii) that all protein contains 16 per cent nitrogen. Protein which is formed from digestion within the rumen supplies the nitrogen requirements of rumen microorganisms. The protein requirements of the grazing animal are supplied from digestible microbial protein, which is synthesized from the protein formed in the rumen and from protein which is not degraded in the rumen. Butler & Bailey (1973) and McDonald, Edwards & Greenhalgh (1975) provide important background to the chemical composition of herbage, and its measurement and expression in relation to animal nutrition.

The intake of herbage is limited by the rate of digestion of digestible material and the rate of passage of undigested material. These parameters combine to determine the extent of digestion, which has commonly been measured as apparent digestibility. It now seems that, despite the view of recent authorities that digestibility is of prime importance in determining intake (Spedding, 1971; Smetham, 1973), we need to appreciate the chemical and physical characteristics of plant cells and how they develop, age and respond to their environment, in order to predict closely the management of grasslands appropriate for maximum animal intake and production.

6.1.1

Cell structure

Cells are composed of soluble material within the cytoplasm and vacuole and of organelles such as the cell nucleus, mitochondria and, in leaves and stems, chloroplasts, 'held in place' by membranes and cell walls. The soluble material is transported round the plant within thick-walled conduits known as vascular bundles. Rigidity is achieved through the laying-down of linear polymers of sugars (cellulose) and cross-linked polymers containing various different types of sugars (hemicellulose) which in turn link with other material, e.g. protein and silica, to form the cell walls; extra rigidity is achieved by amalgams of the carbohydrates and three-dimensional networks of phenylpropane (lignin). Protein within the cells is confined to enzymes ('soluble' protein) and membranes. The complexity of the chemical components within cells may be visualized as shown in Fig. 6.2.

McManus, Robinson & Grout (1977) and others have produced excellent electron micrographs which show that during digestion by grazing herbivores the soluble material is lost rapidly from the plant cell and the cell walls are degraded gradually so that after digestion there remains structure comprised of lignin and some mineral residues which are not digested.

The 'soluble' carbohydrates include disaccharides, the fructans and starch; these are virtually 100 per cent digestible. Likewise, cellulose in its pure form is 100 per cent digestible although it may be less accessible to degradation when it is bound within plant tissues. Hemicellulose, which is bound into the cell wall and protected by coatings of lignin to various extents, has a variable but lower digestibility, as does lignin itself. Isolated hemicellulose has a digestibility of 90 per cent whereas within plant material the digestibility of hemicellulose is 70 per cent and that of lignin is usually less than 3 per cent (Minson, 1982). Cell wall material in young herbage has a digestibility of as much as 90 per cent but this falls to 30 per cent as the cell becomes increasingly lignified with age.

The extent of protein degradation in the rumen is about 90 per cent in early growth and declines to 37–53 per cent in mature herbage (Corbett, 1986). Measurement of degradation of dietary protein in the rumen is, however, subject to large errors, because up to 90 per cent of protein leaving the stomach may be microbial. The apparent digestibility of minerals is highly variable, ranging from 30 per cent for magnesium to 60–90 per cent for phosphorus (Butler, 1973).

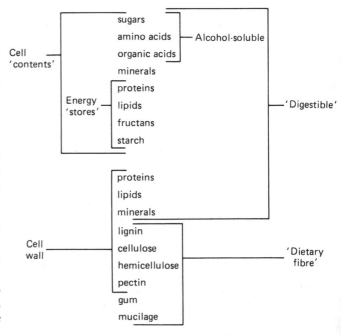

Fig. 6.2. Schema grouping chemical compounds within cells according to alcohol solubility and digestibility by ruminants.

6.1.2

Variation among species

Tropical grasses have the C_4-pathway of photosynthesis and most have a leaf structure in which each vascular bundle is surrounded by a sheath comprised of collenchyma cells which are rich in starch and hemicellulose (thickened walls) and which have the C_3-photosynthetic pathway. Temperate grasses (Festucoideae) and temperate and tropical legumes have only the C_3-pathway and they lack well-developed bundle sheaths. Temperate grasses also mostly accumulate long-chain fructose polymers (fructans or fructosans) rather than starch. Among 25 species, Ojima & Isawa (1968) found almost continuous variation in the relative proportions of fructosans, starch and soluble carbohydrates: many C_3-grasses, e.g. *Agrostis*, *Phalaris* and *Poa* had high levels of fructosans and almost no starch while the reverse was true for legumes, e.g. *Trifolium* spp. and *Medicago sativa*, and C_4-grasses, e.g. *Cynodon dactylon* and *Paspalum notatum*.

The C_3-photosynthetic enzyme Rubisco (ribulose bisphosphate (RuP_2) carboxylase–oxygenase) accounts for about half of the soluble protein in C_3-plants while in C_4-plants, where it is restricted within the bundle sheath cells, it accounts for only 20 per cent of the soluble protein (Bjorkman, Boynton & Berry, 1976).

Chemical and anatomical differences among C_4- and C_3-species are now considered to account for differences in their digestibility (Fig. 6.3). The outer surfaces of the bundle sheaths of C_4-species are not attacked readily by rumen bacteria and may be still relatively intact, and forming a protective shell around the vascular tissue, after 48 h of digestion (Wilson & Hattersley, 1983). By contrast, poorly-developed bundle sheaths which lack suberin, as in C_3-species and C_3/C_4-*Panicum* hybrids, are easily digested by rumen microflora. The C_4-species, in which the ratio of area of mesophyll to area of (less digestible) bundle sheath varies from 1.8 to 3.7:1 (depending on the type of C_4-pathway), have a significantly lower digestibility (Fig. 6.3) than do C_3-species, in which the ratio is approximately 8:1 (Wilson, Brown & Windham, 1983; Wilson & Hattersley, 1983). Likewise, there was a two-fold difference in cell wall content among species in a survey of 13 tropical and 11 temperate grasses (Ford *et al.*, 1979). The cellulose content of tropical grasses and temperate grasses is similar but the tropical species have more hemicellulose than do temperate grasses (Norton, 1982). Legumes have less cell wall than grasses; temperate legumes renowned for high digestibility contain lower concentrations of both cellulose and hemicellulose (Bailey, 1973) and very much

lower concentrations of silica than do grasses (Fig. 6.3). Both tropical legumes and grasses contain a higher percentage of crude fibre (on average 32 per cent) than do temperate grasses and legumes (25 per cent); this suggests an overriding effect of environment rather than differences in composition on quality differences between C_3- and C_4-species (Section 6.1.4).

Legume fibre has a shorter retention time than grass fibre in the rumen. Wilson (1986) suggests that short retention time may be associated with reticulate venation of legume leaves. Reticulate venation results in less vascular tissue per unit volume and many

Fig. 6.3. Tropical (---) and temperate (——) grasses and legumes differ in nutritive value as demonstrated by (a) digestibility, (b) concentration of crude fibre and (c) concentration of silica. Samples comprised more than 600 grasses and 70 legumes; histograms are frequency distributions. (From Wilson & Minson, 1980.)

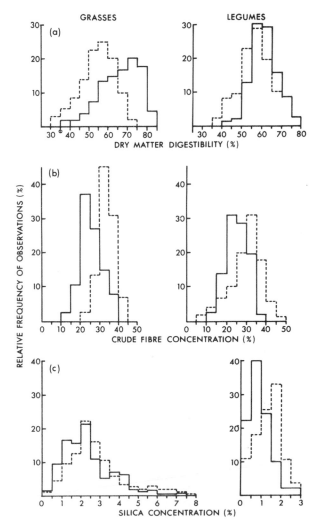

angular junctions between short veins, which may mean the fibre is more easily fragmented into small particles than in the parallel-veined or 'girder' system of grass leaves.

Within major groups of plants, as a generalization there are appreciable differences in the amount and composition of soluble, and thus readily-digestible, material (e.g. Ehara & Tanaka, 1961; Ojima & Isawa, 1968). However, ash (mineral) and fibre content may not vary much within groups. Hunt (1965) found little difference in gross energy value among seven temperate grasses. Despite such generalizations, there are many examples of differences among closely-related species in their acceptability to animals and their degradation within the rumen or under artificial conditions. For example, in a field experiment, giant ragweed was avoided by grazing herbivores despite its having lower concentrations of cell wall and hemicellulose than the preferred grasses (Table 6.1). Among ryegrasses, animal performance was greater and herbage degradation in the rumen faster from *Lolium perenne* × *L. multiflorum* than from *L. perenne*, although their concentrations of structural carbohydrates were similar.

Protein concentrations are higher in C_3-species than in the C_4-tropical grasses (for a review see Hocking *et al.*, 1984). In addition to these differences in protein concentrations there are great differences among species in protein solubility, i.e. the speed and extent to which protein is catabolized within the animal. Differences in solubility may be two-fold within a major pasture group, e.g. 19 per cent soluble protein in *Setaria* compared with 33 per cent in *Brachiaria* within the tropical grass group (Ali & Stobbs, 1980). Low solubility of proteins is correlated with high concentrations of tannins within the feed (McLeod, 1974). Amino acid composition of proteins does not vary greatly among species (Lyttleton, 1973) or with ageing or fertility (Hodgson, 1964; Graswami & Willcox, 1969).

Non-protein nitrogen, which includes nitrate and ammonia, commonly ranges from almost zero to 25 per cent of total nitrogen (say 1.5 per cent of dry matter) within a plant. In a survey of Netherlands grasslands, nitrate averaged 0.64 per cent of dry matter; nitrate poisoning may occur when concentrations exceed 6 per cent (Kemp & Geurink, 1978). Nitrate concentrations vary between species and among plant parts and with environment. For example, at day/night temperature regimes ranging from 15/10 to 33/28°C no nitrate was detected in maize leaves by Muldoon, Wheeler & Pearson (1984) whereas concentrations were as high as 0.4 per cent in millet leaves and nitrate concentrations within stems exceeded 0.6 per cent in maize and millet at day/night temperatures of 18/13 and 21/16°C.

Mineral composition varies among species and it is particularly sensitive to the balance between growth and mineral uptake, i.e. environment and mineral availability. Table 5.2 gives values for mineral concentrations within ryegrass and white clover. Norton (1982) collated mineral concentrations within various pasture groups. He concluded that although the range in phosphorus concentration was similar for tropical and temperate species, few temperate legumes (13 per cent of the survey) had concentrations below 0.24 per cent, which is considered to be the minimum animal requirement, whereas 44 per cent of temperate grasses and 63 per cent of tropical grasses were below this dietary requirement. Legumes have a higher calcium and magnesium concentration than grasses. Furthermore, low calcium may be exacerbated by high oxalate, which renders calcium unavailable. Few data are available for the trace elements (see Section 5.1).

6.1.3
Ageing
As herbage ages the percentage of soluble material within the cells falls and the concentrations of

Table 6.1. *Palatability of rangeland grasses and a lack of relationship between palatability and any simple chemical measurement of the quality of herbage*

	'Palatability'[a]	Digestibility[b]	Cell wall	Hemicellulose[c]	Fibre
Palatable: yellow foxtail	90	74	44	19	27
Palatable: barnyard grass	82	79	49	18	29
Unpalatable: great foxtail	35	72	51	20	31
Unpalatable: giant ragweed	0	72	37	4	32

[a] Palatability was assessed as the percentage of available food which was consumed after 12 days.
[b] Mean percentage digestibility over 3 years.
[c] Percentage hemicellulose.
Source: Marten & Andersen (1975).

hemicellulose, cellulose and lignin rise (Figure 6.4). Van Soest (1967) found cell wall digestibility (CWD, expressed as a decimal) to be related to the degree to which the cells are lignified (L) where the degree of lignification was obtained experimentally as the ratio of lignin to fibre:

$$CWD = 1.47 - 0.79 \log L \qquad (6.4)$$

The ageing, time or stage-of-growth effect is the biggest single variable in herbage quality. It causes digestibility to fall from 75–80 per cent in immature grass or legume to 50 per cent in old pasture in winter in Europe and to as low as 30 per cent in standing dead material in the dry season in the wet-and-dry tropics (Fig. 6.5). Wilson (1982) gave mean rates of decline in digestibility once leaves aged beyond reaching their full area, of 1.5, 1.1 and 1.3 per cent digestibility per week in green panic, buffel and spear grass respectively. More generally (Vickery & Hedges, 1972):

$$D = 0.75 + 0.15 \exp(-0.2t) \qquad (6.5)$$

where D is digestibility and t is time (weeks after germination of an annual grassland). Likewise, the digestibility of crop residues and by-products varies from almost zero to 80 per cent, largely because of variation in their cell wall content (Fig. 6.5).

The rise in lignin concentration and decline in digestibility are most important as pasture reaches flower emergence. Prior to this time the pasture is growing most rapidly so that ageing of old organs is offset partially by new growth. Hay-making commonly takes place when the inflorescence is present but immature. Lucerne (*Medicago sativa*) is used as a pasture or cut for hay; when used for hay it is recom-

Table 6.2. *Percentage composition and digestibility of dry matter of lucerne at stages of flowering when it is often cut for hay*

	Pre-bud	1/10 flower	Full flower
Fibre	22.1	26.5	29.4
Indigestible fibre	8.0	12.8	16.2
Crude protein	25.3	21.5	18.2
Mineral	12.1	9.5	9.8
Digestibility	73	65	56

Source: Watson & Nash (1960), except digestibility from C. J. Pearson, unpublished observations.

mended that it is cut at 'one-tenth bloom' or 'in bud'. Rapid, deleterious changes in chemical composition occur in lucerne at about the time of hay-making (Table 6.2).

The age-related increase in fibre increases the time the feed stays within the rumen of the grazing animal. The relation between intake and retention time is:

$$I_A = 1950 - 51.1t_r \qquad (r^2 = 0.93) \qquad (6.6)$$

where I_A is the intake in kg per sheep per day and t_r is the retention time in the rumen in h (Thornton & Minson, 1973). Thus, increased fibre causes decreased intake irrespective of other changes in chemical composition. For example, the likely intake of a sheep weighing 40 kg may be 1.2 kg per day when fed grass which is low in fibre whereas it falls to 1.0 kg per day for medium fibre and 0.8 kg per day for a diet high in fibre (Hogan, Kenny & Weston, 1987).

Fig. 6.4. Schema of changes in the chemical composition of grasses as they age. (From Osbourne, 1980.)

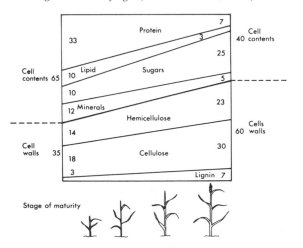

Fig. 6.5. Digestibility of grasses in temperate and tropical grasslands and of grain and by-products in the tropics. (From McDowell, 1985.)

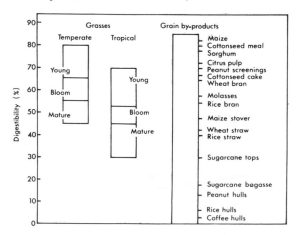

These changes in the chemical composition of cells are amplified by changes in morphology. As herbage ages, the ratio of stem to leaf increases. The stem initially contains high concentrations of soluble carbohydrates and its digestibility may be above that of the leaves (Terry & Tilley, 1964). However, as the stem ages its soluble carbohydrate content decreases more rapidly, and its lignin content increases more rapidly, than those of leaves, so that the decline in digestibility of the feed on offer to the animal is greater than the decline in digestibility of the leaf fraction alone.

The age-related increase in hemicellulose, cellulose and lignin within the ingested feed gives rise to high levels of acetic acid within the rumen of the grazing animal (McDonald *et al.*, 1975). Acetic acid is used with relatively low efficiency by the animal (59 per cent compared with 90–100 per cent for carbohydrates). Thus the decline in herbage quality due to ageing is due both to chemical and morphological changes in the feed and to the way in which the feed is utilized within the animal.

6.1.4
Environment

Temperature is the most important environmental influence on herbage quality (Wilson, 1982). High temperatures accelerate growth, flowering and maturation and in so doing they increase lignification, decrease soluble carbohydrate concentrations and decrease digestibility. Cell wall concentrations increase with increasing temperature (Deinum & Dirven, 1976). The increase is largely in the less-digestible fraction of the cell wall: cell wall digestibility was reduced by 10 per cent when the day/night temperatures were increased from 18/10 to 32/24°C (Moir, Wilson & Blight, 1977). The digestibility of grass tops decreases by about 0.5 percentage units per °C increase in temperature; legumes may be a little less sensitive (Wilson, 1982).

Such effects of temperature may not, however, be of overriding usefulness: Wilson (1982) cautioned that beneficial effects of low temperature on herbage quality in cooler seasons may not always be evident because slower growth at these times, especially of tropical species, usually means that even the youngest herbage is relatively mature and of low dry matter digestibility.

Effects of low radiation, inadequate or excess water, etc. may, like temperature, be explained in terms of environmental influences on plant carbon and nitrogen metabolism and structure. Shading or cloudy weather might be expected to reduce soluble carbohydrate concentrations and thereby increase the percentage of the low digestibility material in the feed. Sheep fed on perennial ryegrass (*Lolium perenne*) shaded to one-quarter of sunlight ate 9–15 per cent less and had a 38 per cent lower gain in liveweight than sheep fed on grass grown in full sunlight (Hight *et al.*, 1968). There are few data on waterlogging: a lower yield would be expected but protein and digestibility may not change in over-wet conditions (Peterschmidt, Delaney & Greene, 1979). Water deficits, insofar as they retard cell ageing and the deposition of cell walls and lignin, reduce by one-third to one-half the rate of the decline with age in the digestibility of grasses (Wilson, 1983). The digestibility of tropical legume leaves, at least the legume Siratro, may be less sensitive to environment than are grasses. Siratro is not affected by shading (Wilson & Wong, 1982) or water deficits (Wilson, 1983) perhaps in part because Siratro does not accumulate highly digestible solutes when it is stressed (Wilson *et al.*, 1980).

6.2
Injurious substances

Herbage plants and weeds which occur within grasslands may be harmful if, by being eaten or touched, they reduce animal productivity. Reduced productivity may simply be reduced intake of feed or less efficient conversion of feed into animal product; it may also involve clinical disorders (poisoning) and even death. Injurious substances within grasslands fall into three categories:

(i) Inorganic compounds and minerals: nitrate, copper, molybdenum, selenium and potassium.
(ii) Organic compounds within plants: diverse chemicals appear to have evolved to protect plants from predation, particularly from insects (Mooney, 1972; Kingsbury, 1983). Some injurious compounds are listed according to chemical type in Table 6.3
(iii) Fungal and microbial toxins.

Injurious substances are widespread. It is estimated that 7000 species contain injurious substances (Culvenor, 1970) and annual losses due to loss of livestock production exceed A$80 million in Australia alone (Culvenor, 1987). Production is lost due to feed which is not utilized for fear of harming the animals, decreased intake and growth, decreased fertility, decreased life expectancy and death, and the cost of disease control measures. Death of stock accounts for 10–15 per cent of the total costs. The diverse toxins and their effects on livestock are described in some detail by McBarron (1976) and Everist (1981); the toxins of most importance among the temperate species *Lolium*, *Phalaris* and *Trifolium* are reviewed by Culvenor (1973, 1987) and Lloyd-Davies (1987).

Table 6.3. *Some injurious organic compounds which are widely found in plants grazed by animals*

Chemical group	Compound	Main pasture plants affected	Animal disorder
Alcohols	Tremetol	Ryegrass, *Eupatorium rugosum*	Tremors, convulsion
	Acetylenic alcohols	Umbelliferous spp., e.g. *Oenanthe crocata*	Tremors, convulsion
Glycosides	Cyanogenic (nitrile) glycosides, e.g. dhurrin, limarin	Sorghum, white clover, cycads (palms)	Anoxia: staggering, death
	Azoxyglycosides, e.g. tycasin		
	Cardiac glycosides, e.g. oleandrin	*Nerium* (oleander)	Anoxia: staggering, death
Oxalic acids	Metal salts and acids	*Oxalis, Rumex* (dock), Chenopods (saltbushes)	Low calcium, low milk, anaemia
Monocarboxylic acids	Fermentable lactic acid	Grain having high carbohydrates	Over-eating
	Betenolide	*Fusarium* – nycotoxin within tall fescue	'Fescue foot'
Amino acids	Mimosine	Leucaena, mimosa	Poor liveweight gain
	Indospicine	*Indigofera*	Poor liveweight gain, abortion
Proteins	Thiaminase	Bracket fern (*Pteridium*)	Monogastrics (e.g. horse) have uncoordination tremors, death
Aliphatics	Glucosinolates (thioglucosides)	Brassicas (cabbage, rape, etc.)	Goitre
		Radish, mustard	Goitre
	Allyl polysulfides	Onions	Anemia; flavour in milk
Alicyclics	Terpenes	Camphor, larkspurs	Multiple symptoms, sometimes death
	Triterpenes	Various, including cucurbits (melons)	Photosensitization, paralysis, death
	Saponins	Legumes, e.g. lucerne	Poor liveweight gain; a factor in bloat
Coumarins	Dicoumarol	*Anthoxanthum* (vernal grass), *Melilotus* spp.	Haemorrhaging, sedation, death
Isoflavenoids	Formononetin	Subterranean clover, red clover	Infertility, difficult lambing
Alkaloids	Mimosine	*Leucaena leucocephala*	Hair loss, infertility
	Coniine	*Conium* (hemlock)	Paralysis, death
	Pyrrolizidine alkaloids, e.g. heliotrine	Wide range, e.g. *Echium, Heleotropium, Senecio, Crotolaria*	Liver lesions, death
	Quinolizidine alkaloids, e.g. lupinine	*Lupinus* sp. (lupin) *Phomopsis* mycotoxin	Paralysis, convulsions, death
	Indole alkaloids, e.g. tryptamine alkaloid	Phalaris, ryegrass ergot (*Claviceps*) mycotoxin	Paralysis, convulsions, death
	Glycoalkaloids, e.g. solanidine	Green *Solanum tuberosum* (potato)	Paralysis, convulsions, death
	Diterpenoid alkaloids, e.g. delphinine	*Delphinium* (larkspurs)	Paralysis, convulsions, death

The concentrations of toxic compounds vary within species. It has thus been possible to select and breed cultivars having low or nil levels. For example, alkaloid concentrations in strains of *Phalaris aquatica* vary from 0.05 to 1.8 mg per g (Oram & Williams, 1976) and *Trifolium subterraneum* cultivars contain concentrations of formononetin which vary from 0 to 20 mg per g.

However, increasingly, animal disorders are being identified as caused by both compounds within the plant and substances produced by bacteria or fungi which are parasitic on the plant or within it. For example, 'ryegrass staggers' is a major problem of cattle and sheep grazing on perennial ryegrass (*Lolium perenne*) in New Zealand and other cool temperate grasslands. Staggers is most likely caused by lolitrems. These are compounds produced by an endophytic fungus, *Acremonium loliae* (Gallagher, White & Mortimer, 1981; Mortimer & Menna, 1985); it is not known whether the lolitrem is synthesized by the fungus or by the plant in response to the fungus. Closely-related toxins are also produced by *Penicillium* in the topsoil of ryegrass pastures and the grass itself produces alkaloids (Culvenor, 1987). Unfortunately, the endophyte *A. liliae* protects ryegrass against Argentine stem weevil and in some way stimulates growth relative to uninfected plants (Latch, Hunt & Musgrave, 1985). Other fungal endophytes, *Acremonium coenophialum* and *Balansia* spp., infect tall fescue (*Festuca arundinacea*). Again, the fungi are carried in the seed so that they are widespread in temperate grasslands; Bacon *et al.* (1986) have shown that one species of parasitic fungus, *Balansia epichloe*, has a wide range of grass hosts and geographic occurrence. This may cause a range of animal disorders generally described in the United States, such as fescue foot, fat necrosis and summer syndrome or poor performance (Stuedemann *et al.*, 1986).

Naturally, the concentrations of injurious substances and the rate at which the contaminated feed is eaten are of prime importance in determining the extent to which these substances reduce animal productivity. This is shown in Table 6.4, which shows that the long-term fertility of sheep grazing monospecific swards of *T. subterraneum* was inversely related to the genetically-determined concentration of formononetin in the clover.

The extent of injury is also affected by ecological factors and management. Isoflavone concentrations in *T. subterraneum* increase four-fold under phosphate deficiency (Rossiter & Beck, 1966) and sulphur deficiency (Rossiter & Barrow, 1972) and under waterlogging or water deficiency (Francis & Devitt, 1969). By contrast, in forage sorghum, glycosides which are hydrolysed to cyanide when plant cells are broken on ingestion are partially detoxified if there are high levels of sulphur in the feed (Wheeler *et al.*, 1975). Cobalt administered as a pellet to animals overcomes chronic 'phalaris staggers' (Lee *et al.*, 1957).

Bloat, an over-distension of the rumen and reticulum, occurs in ruminants grazing legumes because of accumulation of a stable, persistent foam. It is widespread to the extent that farmers may avoid clover-based pastures. Some degree of control, which is not totally effective, is achieved by spraying grasslands with oils and surfactants which have anti-foaming properties (Lloyd-Davies, 1987). Breeding or the selection of plants high in condensed tannin or with higher levels of epicuticular waxes offer other control strategies (Rumbaugh, 1985). The sowing of specially-selected non-injurious cultivars is considered in Section 6.6.

Table 6.4. *High-formononetin cultivars of subterranean clover drastically reduce fertility, but lambing percentages of ewes grazing on low-formononetin clover do not change with time*

Cultivar	Formononetin (% of dry matter)	Lambing percentage		
		After 1 year	After 4 years	After 6 years
Dwalganup	High (1–1.3)	89	6	0
Yarloop	High	82	45	26
Daliak	Low (0.1–0.2)	95	93	77
Seaton Park	Low	88	87	80
Control	Zero	91	59	85

Source: CSIRO (1975).

The toxic effects of mimosine in *Leucaena leucocephala*, a tropical shrub used for browsing, have been eliminated by introducing microorganisms which metabolize mimosine in the rumen; these microorganisms survive and spread readily (Jones, 1985).

Notwithstanding these ways of minimizing the effects of injurious substances, the safest management of animals on grasslands containing injurious substances is still to ensure that the herbage is grazed when at its least toxic stage, when the animals have partially-filled stomachs and in conjunction with other feed so that the diet is mixed and the toxins diluted.

6.3
Sward structure

Animal intake is affected by sward height, leafiness, density and distribution. It is affected in two ways. Firstly, the rate of intake is higher when the herbage is high, leafy and dense, and secondly, animals will select food that they can eat more quickly, i.e. differences in sward structure lead to selective grazing. However, sward characteristics which maximize intake differ substantially between animal species because they eat in different ways. Sheep do not extend their tongue and can only bite up to 4 cm width at a time (depending on age), whereas with cattle, the intake per bite is determined by the size of the arc swept by the tongue during prehension.

Increases in both height and density (number of tillers per m^2) of artificially-prepared grass swards lead to increases in intake by increasing the amount of feed removed in each bite from 0 to 200 mg per bite for sheep (Black & Kenney, 1984). The feed removed per bite increased linearly with sward height at least up to 30 cm (Allden & Whittaker, 1970). This increase with height is not necessarily related to changes in the density of the sward with height (Hodgson, 1981). It assists in explaining why animal intake varies seasonally: irrespective of the amount of available feed, intake is highest on erect spring swards in England (Hodgson, Capriles & Fenlon, 1977). In addition to variation in intake with height of sward, the bite size of cattle grazing tropical grasslands, which are usually tall and sparse (few tillers per m^2), increases with increasing density and leafiness (leaf per unit of height) (Stobbs, 1973*b*). Thus, nitrogen fertilizer was found by Stobbs (1973*a*) to increase the bite size of cattle by increasing leafiness in the early growth of setaria (*Setaria sphacelata*) and rhodes grass (*Chloris gayana*) whereas fertilization reduced the bite size on mature swards because it caused the grass to have higher stem and inflorescence contents and less accessible leaf. In other experiments Kenney & Black (1984)

found that the rate of intake doubled as the length of the feed material (in this case, chopped hay) decreased from 30 to 4 mm.

From these experiments we may conclude that, other things such as sensory factors being equal, the intake of sheep and cattle will be highest from herbage which is tall, leafy and dense relative to the animal's bite characteristics and which fragments easily into small pieces. Grazing experiments confirm that as grazing time increases and the amount of feed decreases there is a decline in the intake per bite, which may be more severe for cattle than for sheep, and a trend to an increased amount of time spent grazing (e.g. Forbes & Hodgson, 1985). However, whereas dense tropical swards allow greater intake than do open swards, in the case of temperate grasslands tiller density may have little effect on animal intake (Forbes & Hodgson, 1985). Moreover, there is a limit to which animals can compensate for reduced intake per bite: Stobbs (1973*a*) and Jamieson & Hodgson (1979) have found that cattle and sheep rarely take more than 36 000 bites per day irrespective of bite size.

Sward structure and the proportion of green, as opposed to dead, material are the two factors most responsible for diet selection; only sensory factors (taste, odour and texture), which we cannot now quantify, may be of equal importance (Black, Kenney & Colebrook, 1987). Animals select feed which they can eat rapidly (Kenney & Black, 1984): they discriminate on the basis of rate of intake between plants or plant parts although sheep have difficulty discriminating between parts that are within 20 mm of one another (Black *et al.*, 1987). The importance of particle size continues into the rumen: Wilson (1986) suggests that the rate of movement of feed out of the rumen is the result of a fluid turbulence process not unlike the movement of sand in water. The particle size with the highest probability of being removed by turbulent flow is 0.23 mm.

The structural strength of leaves or stems, the force required for their removal and the number per bite also affect diet selection. Research into stiffness, tensile strength and fracture properties of fibre bundles is attempting to relate physical aspects of the diet to selectivity (Wilson, 1986; Hodgson, 1986).

6.4
Herbage availability: grazing pressure

Grazing pressure relates the stocking rate (N, animals per ha) to the amount of available herbage or, by another name, the feed on offer (W, kg per ha). The units of grazing pressure may be animals per kg or kg animal biomass per kg available herbage. Equally, for management, animal numbers may be converted into

3 of feed likely to be eaten by the animals. For example, the intake of dry matter by lactating Friesian cows averages 12 kg per cow per day or 27 g per kg liveweight (LW) per day (collated from ten field experiments in England and Australia). Therefore, potential intake relative to available herbage for a Friesian herd of N cows may be written as:

$$P' = 12N/W \qquad (6.7)$$
$$= 0.027 \times N \times LW/W \qquad (6.8)$$

where P' is the potential intake per available herbage, a dimensionless quantity which changes from day to day as the herbage on offer increases because of growth and is reduced due to removal by grazing.

When P' is at or above unity the animals have no opportunity for selecting their diet, only for rejecting some feed and in so doing reducing their intake. The acceptability of herbage plants and browse shrubs varies between species. When P' is at about unity there do not appear to be widespread differences in the acceptability of sown temperate species (e.g. Arnold, 1964) although there are clear differences, which reduce intake, in the acceptability of native species (Robards, Leigh & Mulham, 1967).

When P' is reduced below unity there occur changes in animal behaviour, intake and productivity. This is illustrated by a study of lactating Friesian cows grazing on perennial ryegrass in England. Here, decreasing P' below unity causes increases in grazing time and the number of bites per day and changes in behaviour: the number and length of grazing periods (Table 6.5). Consequently, as grazing pressure decreased from 1.33 to 1, 0.6 and 0.4, i.e. potential intake

per available herbage fell from 1.1 to 0.36, daily intake increased from 11 to 14 kg per cow respectively. These changes occurred in addition to seasonal changes in intake and grazing behaviour (Table 6.5).

Animal intake increases asymptotically with decreased grazing pressure or increased available herbage (Fig. 6.6a). This results from changes in grazing behaviour, as mentioned above, largely because decreased grazing pressure allows increased diet selection. Greater diet selection causes an increase in the quality of feed entering the animal, which in turn permits more rapid digestion and passage of the feed through the animal. This is illustrated in Fig. 6.6b. An increase in the amount of available herbage causes the digestibility of the green feed on offer to decline because it is comprised of an increasing proportion of old leaves and stems of relatively low digestibility; however, the digestibility of all the feed entering the animal's oesophagus increases because the animal can avoid dead material and become increasingly selective about the quality of green herbage which it eats. The percentage of green material in the intake by sheep of four temperate perennial grasses increased with the amount of food on offer in an asymptotic manner:

$$g = 95[1 - \exp(-0.0022W)] \quad (r^2 = 0.85) \qquad (6.9)$$

where g is the percentage of green feed and W ranged from zero to 2500 kg of feed per ha (Hamilton *et al.*, 1973).

It is generally accepted that effects of management such as grazing pressure on intake are mediated largely through changes in the digestibility of the diet. We might expect a positive curvilinear relationship

Table 6.5. *Effects of grazing pressure and season on grazing behaviour of Friesian cows grazing on* Lolium perenne *in England*

	Potential intake per available herbage[a]			Season[a]		
	1.1	0.54	0.36	May	June	July
Organic matter intake (kg per animal)[b]	11	13	14	16	12	10
Grazing time (h)	7.6	8.7	8.8	7.7	9.0	9.4
Total daily bites ($\times 10^3$)	28	34	34	30	35	32
Bite size (g/bite)	0.39	0.40	0.43	0.56	0.37	0.30
No. of grazings/day	5.4	6.1	6.5	6.7	6.0	5.2
First grazing period (h)	13.7	3.5	2.7	2.1	3.6	4.3

[a] Values for potential intake relative to available herbage are means over season and values for season are means over all grazing pressures.
[b] All measurements except bite size differed statistically between grazing pressure treatments and seasons.
Source: Le Du *et al.* (1979).

Fig. 6.6. (a) Herbage intake increases to an asymptote with increase in feed allowance or decrease in potential intake relative to allowance, due in large part to (b) greater diet selection and thus high quality of intake. In (a) the intake by dairy cows is scaled relative to the maximum intake in each of five experiments in the United Kingdom; (b) refers to the diet of sheep in Australia. ((a) Adapted from le Du, Combelas, Hodgson & Baker, 1979; (b) from Hamilton, Hutchinson, Annis & Donnelly, 1973.)

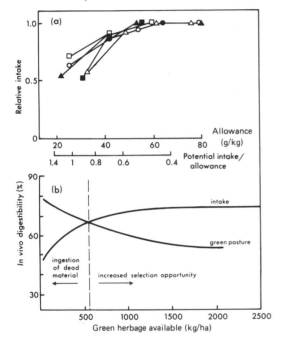

between intake and the digestibility of the feed entering the animal. However, as shown in Fig. 6.7, which summarizes much of the data collated for temperate species, intake increases almost linearly between digestibility values of 55 and 80 per cent, which is the range commonly found in extensive grasslands. The data shown in Fig. 6.7 and from other studies (e.g. Milford & Minson, 1966 for nine tropical species) also show that, depending on the pasture species and grazing conditions, there are large differences in intake of different herbage which have the same digestibility (see Section 6.1.2).

6.5
Animal type and productivity

It is not possible to determine the value of feed for animal productivity without reference to the animal(s) for which it is intended. Ideally information should be available on:

  (i) the likely intake of feed;
 (ii) the potential yield of absorbed nutrients; and
(iii) the potential yield of absorbed energy or
       data from which such information can be deduced
       (Graham, 1983).

Where forage is the sole source of feed for grazing ruminants it must provide the rumen with all the nutrients essential for efficient fermentative digestion. Proteins and starch which 'escape' through the rumen undegraded provide energy for any demanding physiological activity, e.g. pregnancy and lactation (Fig. 6.8).

Fig. 6.7. Relationship between intake and digestibility for various temperate grasses and legumes. Intake is related to body weight raised to the power 0.75 to relate it to 'metabolic weight', i.e. to minimize differences among livestock of different sizes. (From Minson, 1982, based on data of Demarquilly & Weiss, 1970.)

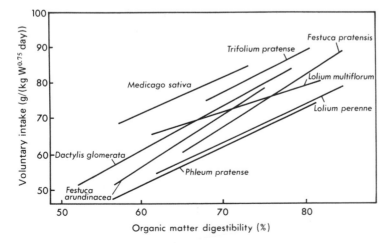

6.5.1
Animal type

The type, weight, physiological state and productivity of the grazing animal are important to intake (e.g. Weston, 1982). Animals differ in their physical capacity to eat (bite size, grazing time, rumen volume, etc.), in their ability to utilize a given feed and in their requirements for feed. Of the variables affecting intake, digestibility can vary over a two-fold range, the efficiency of metabolism over a less than two-fold range and intake over a five-fold range (Egan *et al.*, 1986). Differences between animal types which influence feed intake are thus of significance.

Details of the eating characteristics of livestock are beyond the scope of a book on agronomy. Nonetheless, differences among animals in their ability to extract nutrients from pasture are interesting and of potential importance in management. For example, sheep, with smaller bite size and lips which can move away from the teeth, graze closer to the ground than cattle (Dudzinski & Arnold, 1973). Sheep are thus often grazed in fields after cattle have been removed to obtain greater utilization of available herbage. Goats may preferentially select against legume in legume–grass mixtures and often preferentially select woody herbs. They may thus utilize feed which is rejected by other livestock. Sheep and goats together may thus more effectively manage botanical composition and utilize grass–legume pastures than would either species alone.

Diet selection, as in the case of sheep, cattle and goats, may be more responsible for reported differences in apparent digestibility than are differences in the animals' ability to extract nutrients from the herbage within the rumen. Taylor (1986) highlights the failure of research on intake to allow for animal behaviour patterns and to make measurements that are appropriate.

Intake of feed of the same measured qualities (e.g. digestibility) may vary between animal types. For example, intake, without the potential for selection, varies between contrasting animals which co-habit in Australian rangelands. The intake per unit of metabolic body weight ($LW^{0.75}$: Kleiber, 1961) of lucerne hay and oat straw by red and grey kangaroos was compared with that of sheep (Forbes & Tribe, 1970). Intake was higher for the sheep. The length of time the feed remained in the animal was longer for the sheep than for the kangaroos, e.g. 70 compared with 44 h retention of straw, and the sheep digested the straw far better than did the kangaroos mainly because of a greater ability to digest fibre. This emphasizes the need to measure forage quality with respect to the livestock likely to graze the feed.

6.5.2
Utilization of energy

Digestion of carbohydrates, lipids and protein releases energy in the form of chemical compounds which may be used by the animal or used by the microbial population of the gut. End products of digestion include volatile fatty acids, lactic acid, carbon dioxide, methane, glucose and amino acids (Fig. 6.8). The oxidation of these compounds produces heat. Measurement of heat is used to assess the energy which may be extracted from the feed; the units are Joules per weight of feed or more usually Joules $\times 10^6$ or MJ. Animals do not use all the energy in the feed that is eaten; a schematic representation of the recognized partition of energy is given in Fig. 6.9.

Animals use net energy in different ways; when feed only maintains body liveweight the net energy of the feed is equal to the net energy for maintenance (NE_m). In addition there may be net energy for production (NE_p) which can be used for growth, lactation, etc. as shown in Fig. 6.9. Because the relationship for the efficiency of conversion of metabolizable energy to net energy is known for each productive state, feed analyses may be based on measurement of metabolizable energy alone. Feed requirements can then be calculated from the metabolizable energy content of the feed and the animal's metabolizable energy requirement. The concentration of metabolizable energy in a feed source is MD MJ per kg DM (Section 7.4). Armstrong (1982), Graham (1983) and Alderman (1983) review the energy value of herbage, the energy requirements of animals and the energy-based systems of feed analysis. Leng (1986) points out, how-

Fig. 6.8. Schema of digestion in the rumen of a basic diet supplemented with by-pass nutrients. The intake of fermentable material in herbage is affected by the amount of escape or by-pass protein and the roughage characteristics of the diet whereas the efficiency of feed conversion is influenced by by-pass starch. (From Preston, 1984.)

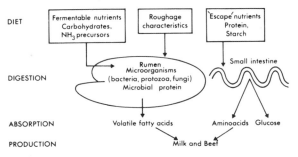

ever, that conversion of metabolizable energy to net energy depends on misleading assumptions of the efficiency of use of metabolizable energy for maintenance and production. These efficiencies are variable and appear to depend on the proportion of metabolizable energy absorbed as glucogenic energy, long chain fatty acids and essential amino acids. The metabolizable energy system of feed analysis has utility in formulating rations during drought (Oddy, 1983) and in feed year planning (Chapter 7).

Leng (1987) suggests the following desirable minimum quality requirements from grassland for maintenance plus production:

(i) The digestibility of feed should be high, i.e. in excess of 65 per cent.

(ii) Soluble protein in the diet should provide about 30 g N per kg digestible dry matter.

(iii) 'Escape' dietary protein should provide about the same protein as arises from rumen microorganisms, i.e. about 200 g protein per kg digestible dry matter intake.

(iv) The level of fat (lipids) in the diet should approach that of temperate grasslands, i.e. 4–8 per cent of dry matter.

(v) Glucogenic energy should be a high proportion of the volatile fatty acid energy absorbed from the rumen, i.e. greater than 25 per cent.

Fig. 6.9. The fate, or partitioning, of the energy within feed, according to its utilization by animals.

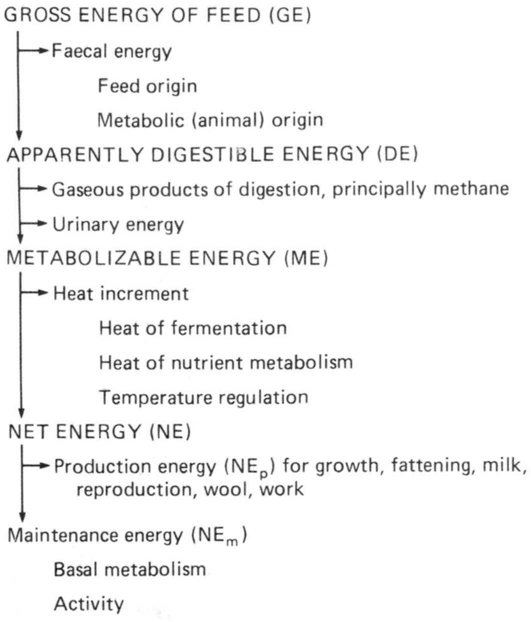

GROSS ENERGY OF FEED (GE)

→ Faecal energy

    Feed origin

    Metabolic (animal) origin

APPARENTLY DIGESTIBLE ENERGY (DE)

→ Gaseous products of digestion, principally methane

→ Urinary energy

METABOLIZABLE ENERGY (ME)

→ Heat increment

    Heat of fermentation

    Heat of nutrient metabolism

    Temperature regulation

NET ENERGY (NE)

→ Production energy (NE_p) for growth, fattening, milk, reproduction, wool, work

Maintenance energy (NE_m)

    Basal metabolism

    Activity

6.5.3
Intake and productivity

Within a particular type or class of animal, intake depends primarily on the size and physiological state of an animal: it needs food to maintain its resting body functions and to provide energy for walking, etc. and for production, e.g. of milk. The intake required for resting functions or maintenance is directly proportional to metabolic weight ($LW^{0.75}$) (Kleiber, 1961). To account for the energy expended in grazing Jagusch (1973) inflated the maintenance requirement by 30 per cent. Some experiments suggest that grazing animals need a 10–50 per cent higher intake than the maintenance level indicated in classical feeding standards (e.g. ARC, 1965): an English study of the energy needs of lambs indicated that, while grazing, they used 80 units of energy (KJ per kg liveweight per h) to eat, 57 units to stand, 20 units to ruminate and only 13 units to walk (Spedding, 1984a).

When an animal is productive, its requirements for energy protein and minerals are raised further. For example, Preston (1972) described the protein intake which is required by cattle in terms of their metabolic weight ($LW^{0.75}$) and daily liveweight gain (G, kg per day):

$$Ip/LW^{0.75} = 1.6 + 5.2G \qquad (6.10)$$

where $Ip/LW^{0.75}$ is in g digestible protein per kg metabolic weight. When productive (e.g. milking) appetite, body weight and the level of production (milk yield) change continuously according to physiological factors such as hormone levels and to season (e.g. Haresign, 1981). For example, seasonal changes in resting metabolic rate, inactivity and catabolism of body fat cause caribou in Alaska to use 45 per cent less energy in winter than in summer (Boertje, 1985).

Growth is at the 'end of the line' as far as the animal's available energy is concerned: liveweight gain or wool or milk production depend on ingestion of sufficient high-quality feed to provide energy in excess of the animal's requirements for maintenance and activity. Thus, although in Section 6.4 we generalized that a lactating cow needed a daily intake of 12 kg of high quality dry matter (say, having an energy concentration of 12.5 MJ per kg), it is more accurate to say that we would expect a 450 kg cow eating 150 MJ per day to produce 15 kg of milk and that milk production will decline or stop if the quality of the feed is poorer or animal liveweight is greater than in our average situation. Stated another way, the intake required for maintenance and production of 30 kg of milk is four times that for maintenance alone (Table 6.6). Cattle being fattened to sustain a liveweight gain of 1.25 kg per day and lambs being fattened to sustain a gain of 200 g per day need approximately twice the intake they need for maintenance.

6.6
Implications for management
Herbage quality and animal intake are closely linked, and animal intake is closely linked with production. It follows that production will be maximized by management that produces a best compromise of matching grassland growth and quality and animal intake. Thus, the goals of management are to:
  (i) minimize the seasonality of available feed or match the pattern of availability with the pattern of intake;
 (ii) maintain a high quality of feed, i.e. a desired chemical composition and freedom from injurious substances;
(iii) maintain a desirable sward structure, particularly with respect to light interception, height, leafiness and density;
 (iv) optimize grazing management with respect to the phenology of the grassland species to encourage seed production and persistence of desired species and reduce or eliminate unwanted species.
  (v) optimize herd structure and operations, e.g. calving and culling, to take advantage of the seasonal availability of feed.

In the present chapter we have considered briefly the feed requirements of individual animals; Section 7.2 deals with herd requirements and production and how these may be manipulated through changes in the stocking rate, herd composition, time of mating and weaning, and grazing management including the extent to which grazing is rotated among fields. Consequently we consider below only management of the

Table 6.6. *Metabolizable energy requirement (MJ/day) of a dairy cow weighing 500 kg and producing milk which has 3.8% fat*

Feed quality (MJ/kg DM)	Milk yield (kg/day)			
	0	10	20	30
9.1	65	136	—	—
10.8	62	125	199	—
12.5	60	122	189	261

Source: Jagusch (1973).

feed supply ((i)–(iv) above) to give optimum animal production.

First, the seasonality of grassland growth is an unavoidable consequence of plant responses to a non-optimum environment (Chapter 3). In most climates, feed is in shortest supply during autumn and winter. In cool temperate climates this is because temperatures are too low for maximum growth (Fig. 3.10); in mediterranean climates pastures, e.g. annual ryegrass and subterranean clover, must regenerate from seed in the autumn in order to produce approximately 80 per cent of the yearly digestible dry matter over a period of 6–10 weeks in spring; in tropical environments the composition of grasslands is dominated by tropical (C_4) species whose growth becomes rank late in the wet season, over-shading species which may be more adapted climatically to growth in the cool or dry season.

The seasonality of grassland growth can be minimized by (i) sowing species which have relatively long growing seasons or (ii) managing the grassland to retain a mixture of species. Alternatively, the farmer may accept the seasonality of feed production and choose instead either to manage herd intake so that, as closely as possible, intake follows the seasonal pattern of feed availability, or to conserve feed as hay or silage (Section 7.3).

Selecting species with long life cycles so that they extend the growing season means replacing annuals with other annuals which have a longer vegetative phase (later flowering) or an indeterminate pattern of growth (flowering and concomitant vegetative growth), or, alternatively, by replacing annuals with perennials. Late flowering may retain high leaf/stem ratios during the growing season, which is advantageous. Minson *et al.* (1964) showed that among temperate grasses, species which delayed flowering or did not flower were more nutritious and of greater value than early-flowering species for animal production. However, late flowering will delay the onset of higher growth rates associated with reproductive swards (Sections 3.3, 4.5) to a period when feed may not be in short supply as say, in late spring.

Indeterminate growth can be advantageous if the inflorescences are highly nutritious, because here quality is maintained at an acceptable level during a prolonged period of flowering. This is found in the case of prostrate legumes, e.g. subterranean clover, less so for erect, stemmy legumes, e.g. lucerne; it also holds for some grasses, e.g. bromus (*Bromus catharticus*), but for many grasses, e.g. ryegrass, flowering and seed production are associated with an unacceptable reduction in digestibility and palatability. Likewise, replacement of annual species with perennials has consequences which are not wholly desirable. Even in climates which are sufficiently mild to permit perennials to persist, the choice of annual or perennial should be based firstly on which species will intercept most radiation throughout the year. A second aspect is that rapid early vegetative growth of annuals (Hill & Pearson, 1985) may offset the perennials' advantage of regrowth from the parent plant at the beginning of the season.

Managing grasslands to retain a mixture of species with different environmental requirements and tolerances is an alternative management strategy to trying to select a monoculture which may produce a seasonal distribution of feed. Mixtures are usually no more productive than monocultures if both species are growing at the one time (e.g. Trenbath, 1978). However, mixtures of C_3- and C_4-species are theoretically desirable if the temperatures approach the optimum for temperate species in one season and tropical species in another. Fig. 6.10a shows the growth rates of C_3- and C_4-species in just such a warm temperate climate. The contribution of various species to the available feed changes radically throughout the year (Fig. 6.10b) although most plants of the main species (*Lolium perenne* and *Paspalum dilatatum*) perennate. Such mixing of species is rewarding if inter-specific competition is minimized and quality is maintained at a high level, with digestibility above 65 per cent. Both these criteria are achieved by heavy grazing, slashing or herbicide application in the autumn to remove the carry-over of standing tropical grass of deteriorating quality.

Maintenance of a feed supply of high quality is achieved theoretically by maintaining the grassland at a young stage, preferably with a high content of legume. This could occur by relatively heavy grazing, so that the animals continuously remove new growth but are not forced to eat the stubble (of poor quality) on which the new growth depends. Such a grazing strategy has advantages in that utilization (feed eaten relative to net photosynthesis) is high (Table 3.3). However, low leaf area indices and high stocking rates may also result in poor interception of light and therefore low grassland growth rates (Section 3.3) and reduced production per animal (Section 7.2).

The maintenance of a significant proportion of legume within the pasture, which is achieved readily at high stocking rates, may result in lower pasture growth rates (the legume growing more slowly than grass: Chapter 3). However, the intake of temperate legumes is invariably greater than that of temperate grasses and this is a major reason why animal production is usually greater on legumes than on grasses. For instance, the intake by cows fed with ryegrass and white clover in stalls was 16.5 and 18.9 kg per day respectively (R. C. Kellaway, 1986, personal communication). The higher intake of clover is associated with a shorter retention time in the rumen. Also, the quantity of protein digested in the small intestine is greater with clover than with ryegrass containing similar

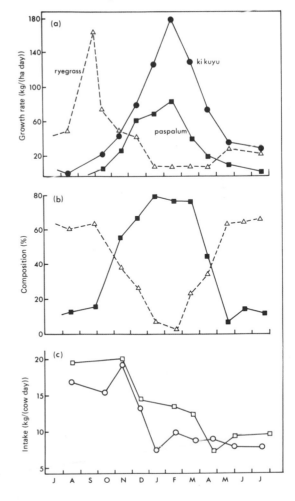

Fig. 6.10. Maintenance and utilization of grasslands comprised of C_3- and C_4-species. (a) Seasonal growth rates of ryegrass (C_3) and of kikuyu (*Pennisetum clandestinum*) and paspalum (*Paspalum dilatatum*) (C_4) when grown in pure swards; (b) seasonal composition of a ryegrass–paspalum grassland; and (c) apparent intake of paspalum–ryegrass in successive years (○, □) by Friesian cows. All the data refer to grasslands near Sydney, Australia. Maximum growth rates and composition changes are related closely to temperature optima (Fig. 3.10) whereas the rapid decline in herbage removal in mid–late summer (December to January) is caused by the declining quality of a grassland then dominated by ageing C_4-species. (From: (a) and (b) C. J. Pearson, 1974–8, unpublished observations; (c) K. Mendra, 1980, personal communication.)

amounts of protein. The legume will also confer advantages to the nitrogen nutrition of the pasture, but these may be offset by the increased acidification it causes (Sections 5.2, 7.2).

We conclude that grassland management for quality should be based firstly on setting a minimum standard of acceptable quality (which may change seasonally depending on animal requirements) and secondly on maximizing grassland growth rates within the quality constraint. How the growth rate (and hence, intake) is maximized will depend on which management options are feasible (Chapter 7) and on economics (Section 8.2). For example, in England, where there is a high level of managerial expertise and prices for animal products are high, it is both feasible and economic to grow ryegrass monocultures using fertilizer nitrogen whereas in Australia and New Zealand legume-based pastures achieve the same quality criterion but lower productivity and without nitrogen, and in temperate Argentina it is generally uneconomic to improve the quality of naturalized temperate grasslands by using either legumes or nitrogenous fertilizers.

Avoidance of injurious substances involves management at the point of species selection (or eradication) and grazing. For some species, e.g. ryegrass, we do not yet know the criteria for selecting cultivars

Fig. 6.11. Competition among cultivars of subterranean clover sown in 50:50 mixtures. (Shaded): Cultivars containing low levels of formononetin; (white): cultivars containing high levels of formononetin. Leaf areas were measured 6 years after sowing. Daliak replaced Dwalganup and Yarloop whereas Uniwager did not compete effectively against Dwalganup. (From CSIRO, 1975.)

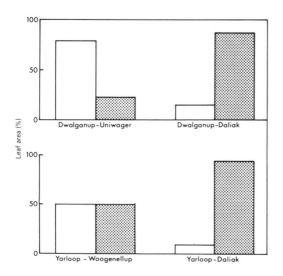

bearing toxic substances or not carrying them. As mentioned above (Section 6.2) the selection of cultivars which are free of toxins may result in lower productivity. By contrast, in subterranean clover and phalaris, new cultivars which carry nil or low levels of toxins are usually agronomically equal to or better than cultivars which contain toxic substances. Thus, prudent management would move to replace old cultivars with new. However, even if the old cultivar is killed using herbicides, the replacement requires that the desirable cultivar is more competitive than the undesirable cultivar, because of the high level of unwanted seed in the soil seed bank. Sowing of binary mixtures shows clearly that competitiveness varies between cultivars so that over-sowing of a grassland may result in anything from virtually complete dominance by the desired cultivar to little change from dominance by the cultivar which contains toxic substances (Fig. 6.11).

Grazing management to avoid the harmful effects of injurious substances is based on the principle that harm is minimized if the concentration of toxin in the ingesta is minimized. This is achieved by grazing when the concentration within the plant is relatively low, mixing the diet, slowing the rate of intake and preconditioning ('hardening') the animals. All these practices usually reduce intake and animal production.

Management can affect sward structure, the third way of easily changing intake, through the selection of grassland species and choice of the intensity of grazing. It is not unequivocal which aspects of structure are most important: a grassland which intercepts all light and is of moderate height, high density and high leafiness is theoretically ideal. This roughly approximates a leaf area index of 3 in legumes and as high as 8 in some grasses, a height of perhaps 30 cm and tightly packed long leaves and pseudostems with no old, true stems which have low quality and are resistant to breakage. We suggest again that light interception is of primary importance because if this is suboptimal, growth will be slow and there will be a higher proportion of old tissue in the sward. Equally, if grazing pressure is too low or the interval between successive grazings is too long, the pasture will contain a high proportion of low-quality material and a relatively large amount will also be lost due to death, pests, diseases and the detritus feeders.

Perennial temperate grasses in the vegetative phase appear to meet the criteria for an ideal sward structure most closely. Also, they tolerate the greatest flexibility (or lack of control) of grazing pressure or grazing interval, insofar as they have low extinction coefficients which will support high rates of growth over the widest range of leaf area indices, from complete ground cover to complete light interception. Currently, farmers choose species on the criteria of production and quality. It is doubtful if differences in sward structure will be shown to have as much effect on animal productivity, or be as simple to manage, as pasture production and quality. Similarly, although grazing pressure may affect the productivity of the pasture, the effect of grazing pressure on net photosynthesis is not great (Vickery, 1972; Parsons *et al.*, 1983*a, b*) and differences in animal intake seem to be explained more by pasture availability and losses (death etc.) than by changes in sward structure.

6.7
Further reading

Butler, G. W. & Bailey, R. W. (eds.) (1973). *Chemistry and Biochemistry of Herbage.* 3 vols. London: Academic Press.

Hacker, J. B. (ed.) (1982). *Nutritional Limits to Animal Production from Pastures.* Farnham Royal: Commonwealth Agricultural Bureaux. Herbage quality and limitations to intake in Parts 3 and 4.

7

Grassland – animal interactions and management

In this chapter we are concerned with interactions between grassland and animals. Animals affect grassland productivity and botanical composition by selective grazing, pulling, treading, aiding seed dispersal (Chapter 2) and the return of nutrients to the ground (Chapter 5) and fouling. The importance of the effects of animals depends on the grazing management system. Grazing management systems are defined as management systems designed to optimize the efficiency of production or total productivity of the livestock (in dollars, product or manager's satisfaction). Thus, grazing management systems attempt to optimize the interaction between grassland and animals given an understanding of grass growth (Chapters 2–4), grass quality, animal intake and animal requirements (Chapter 6). This optimization may involve conservation (Section 7.3), which may be an integral part of feed year planning (Section 7.4). The chapter concludes with an analysis of the efficiency of livestock production.

7.1
Animal effects on grassland

7.1.1
Selective grazing

Given the opportunity, all animals select some plant parts in preference to others and some species rather than others. The basis of this selection, and its consequences for intake, were discussed in Chapter 6. As availability decreases (or grazing pressure increases) selection is reduced (Section 6.4).

Fig. 7.1 shows diet selection by steers grazing a tropical grassland comprised of native grasses and the sown legume *Stylosanthes hamata*. Steers selected green grass early in the wet season but they preferred the legume as the grass aged, flowered and became low in quality; the consumption of legume fell again when it became dry 1–2 months after the end of the wet season.

Sheep are usually more selective than cattle (e.g. Dyne & Heady, 1965).

Selectivity is greatest when animals graze new, unfamiliar pasture at low grazing pressure. In rotational grazing systems, where animals graze a particular field for, say, 2–8 days at increasing grazing pressure before being moved to another field, selectivity is greatest in the first 1–2 days (see Section 7.2.3).

Selectivity results in grazing causing marked changes in the botanical composition and perhaps in the seasonality of growth of the grassland. This is particularly true when grazing affects flowering and seed set (e.g. Heitschmidt *et al.*, 1982). Changes in botanical composition are not necessarily avoided by increasing the grazing pressure and thus reducing selectivity, if this disadvantages only one component of the grassland. For example, the introduction and early heavy stocking of sheep in Australia led to rapid and permanent changes in the composition of semi-arid grassland. In a grassland containing a wide range of species (136 annual and perennial grasses) grazing had no effect on the currently-dominant grass, *Danthonia caespitosa* (Fig. 7.2). This is because the *Danthonia* is a C_3-species which flowers and produces seed in the winter, when 60 per cent of the sheep's diet is made up of other species. By contrast, *Enteropogon acicularis*, a remnant of the original grassland and a C_4-species which grows poorly in winter, flowers in the summer. It is repeatedly defoliated whenever rain falls in summer and its flowering is prohibited. Thus grazing kills *E. acicularis* seedlings and prevents seed production from mature plants, so that age cohorts have a short life expectancy and few seedlings are produced for the next generation (Fig. 7.2).

Fig. 7.1. Diet selection by steers grazing a tropical grassland. Seasonal changes in preference indices (herbage eaten relative to herbage available) for *Stylosanthes hamata* (■) and native grass (○) under low stocking rate in northern Australia. (From Gardener, 1980.)

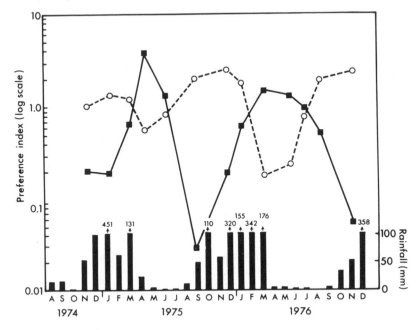

Fig. 7.2. Survival of two native grasses in grazed (□) and protected (■) disclimax grassland in semi-arid Australia. Grazing has no effect on *Danthonia caespitosa* (a) whereas it kills otherwise long-lived *Enteropogon acicularis* (formerly called *Chloris acicularis*) (b). (From Williams, 1970.)

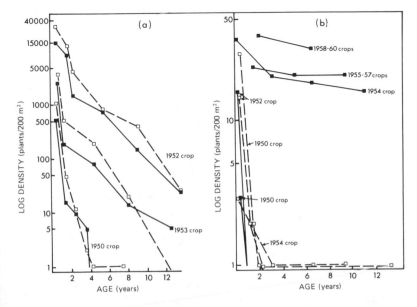

7.1.2
Pulling

Pulling or uprooting of pasture, although recognized as a deleterious aspect of grazing, is thought to cause variable, and usually only negligible, losses in productivity. It is confined to certain times of the year, e.g. mid-summer–autumn in England, and is most severe when grass swards have developed true stems, as occurs during flowering or following lenient grazing. Tallowin (1985) found no effect of pulling on tiller density or ryegrass production one year later.

7.1.3
Treading and poaching

The pressure exerted on a grassland by sheep is estimated to be 0.8–0.95 kg per cm^2, that exerted by cattle to be 1.2–1.6 kg per cm^2 (Spedding, 1971). It has been estimated that grazing animals tread on 0.01 ha per day although, of course, the actual area affected depends on the availability of feed, animal behaviour and the weather (Curll & Wilkins, 1983).

Treading directly damages standing feed and it also has indirect effects, such as soil compaction, which may in turn affect pasture growth (for a review see Watkin & Clements, 1978). The direct effects, termed trampling or poaching of the pasture, may be short term: damaged and buried tillers re-appear within 15 days (Edmond, 1958). Poaching is appreciable when the ground is wet. However, one survey of English farms found that poaching did not markedly reduce grass growth and animal utilization of pasture (which varied among farms from 44 to 70 per cent); utilization did not vary between summer and winter (Peel & Matkin, 1984).

The indirect effects of treading are mostly on soil bulk density and infiltration. At stocking rates of 28 sheep per ha there is a long-term increase in bulk density. The increase in bulk density may be small, e.g. from 1.2 to 1.4 g per cm^3 over two years (Pott & Humphreys, 1983) or appreciable, e.g. a 50 per cent increase from 0.95 to 1.42 g per cm^3 (Curll & Wilkins, 1983) but even in the latter case there was no detectable change in pasture growth or botanical composition. Others have noted increases in soil bulk density (e.g. Haveren, 1983) and reductions in the short-term infiltration of rain (e.g. Gifford & Hawkins, 1978) but it seems likely that treading causes substantial long-term reductions in pasture growth (e.g. Brown and Evans, 1973) only at extremely high intensities of stocking, as occurs in laneways or on susceptible soils.

Witschi & Michalk (1979) and Pott, Humphreys & Hales (1983) carried out elegant experiments in which sheep walked on but did not graze grassland comprising legume and grass. On temperate *Trifolium subterraneum–Lolium rigidum* grassland, treading by up to 39 sheep per ha increased the soil bulk density by 40 per cent and caused a one-third reduction in winter growth due mainly to suppression of the ryegrass (Witschi & Michalk, 1979). By contrast, on tropical *Lotononis bainesii–Digitaria decumbens* grassland, grazing intensities of up to 28 sheep per ha reduced the herbage yield within 30 days whereas 90 days after treading the yield equilibrated because treading stimulated the growth of the grass and suppressed that of the legume. Elimination of the tropical legume was due to the direct effects of grazing causing legume seed reserves to decline from 5800 seeds per m^2 under 7 sheep per ha to 400 seeds per m^2 under 28 sheep per ha. Elimination of the legume was also caused by the indirect effects of treading because the legume was probably more sensitive to compaction of the topsoil than was the grass (Pott & Humphreys, 1983; Pott *et al.*, 1983).

7.1.4
Fouling

Dung and urine cause pasture to be temporarily unpalatable. Fouling thus indirectly causes spatial redistribution of plant species as well as redistribution of nutrients (discussed in Chapter 5). Dung may affect pasture height (and thereby 'rankness') for over 3 years (Jones & Ratcliff, 1983) although on average the effect lasts for 1–1.5 years. Areas near dung pats become dominated by grass whereas grazed, unfouled areas may tend towards legume dominance (Leith, 1960). Hilder (1964) found that sheep camps caused 22 per cent of dung to be returned to less than 3 per cent of the field and Jones & Ratcliff (1983) showed that defecation on such a small percentage of the field nonetheless caused 70 per cent of the land to be grazed lightly. Thus, the major wastage is not rejection of fouled pasture but rejection of pasture surrounding faeces or distant from camps.

Dispersal of the dung by harrowing promotes more even grazing but it does not increase pasture growth (Weeda, 1967).

Animals do not selectively avoid areas of pasture affected by urine unless 'rankness' has resulted (Watkin & Clements, 1978).

The value of dung and urine is that they return nutrients for pasture growth. A single urination by a sheep may be equivalent to one application of fertilizer at the rate of 450 kg N per ha on the urine patch (Doak, 1952). In an early study in New Zealand, the annual grassland yield was higher when nutrients were returned to the grassland than when they were not: 15.4 t per ha compared with 11.7 t per ha (Sears, 1960). This yield response is more striking than is usually found. Nonetheless, changes in botanical composition in the New Zealand study are typical of those reported throughout the world: with return of minerals to the grassland the ratio of grass to legume became 72:26 whereas under lower fertility there was a 50:50 balance between grass and legume.

7.2
Grazing management systems

A role of grassland agronomists is the designing, testing and extension to farmers of grazing management systems. These systems are usually designed to maximize the efficiency of conversion of solar energy or grassland growth into animal product, although other goals may be equally appropriate (Section 8.1). As a generalization, the more highly productive the system, the more complicated its management and the more frequent are proposals for innovation, whereas in extensive grasslands we accept low efficiency and little management. Management systems are thus site- and culture-specific and they change according to economic and technological circum-

stances (Section 8.2). The variables which may be changed are (Spedding, 1975b):

  (i) the plant species, cultivar, age, population;
  (ii) the animal number, breed, size, herd age and age structure;
  (iii) the grazing period;
  (iv) the non-grazing period;
  (v) the grazing method: the frequency of rotation and the size of field;
  (vi) conservation;
  (vii) the supplementary feed;
  (viii) disease control;
  (ix) the infrastructure, including the lay-out of water and feed points;
  (x) the labour requirement;
  (xi) key husbandry practices, e.g. at parturition.

7.2.1
Production per animal and per area

In most cases the average animal production (e.g. liveweight gain per animal, G) declines linearly with increasing stocking rate N. That is (Hart, 1972; Jones & Sandland, 1974; Fig. 7.3a):

$$G = a - (b \times N) \tag{7.1}$$

where G is the gain in liveweight (g per day), a is the intercept i.e. maximum liveweight gain per animal on a particular grassland, which varies from 210 g liveweight gain per day on setaria-siratro and 720 g liveweight gain per day on US rangelands to 1720 g liveweight gain per day on rotationally-grazed ber-

Fig. 7.3. Effect of stocking rate on animal performance, energy use and profitability. (a) Liveweight gain per ha and gain per animal from a wide range of experiments on temperate and tropical grasslands. (b) Liveweight gain and wool growth per ha of sheep on ryegrass–clover in England. (c) Utilized energy output and (d) profit and risk predicted from various rates of stocking of dairy cows in England. (From: (a) Jones & Sandland, 1974; (b) Curll et al., 1985; (c) and (d) Doyle & Lazenby, 1984.)

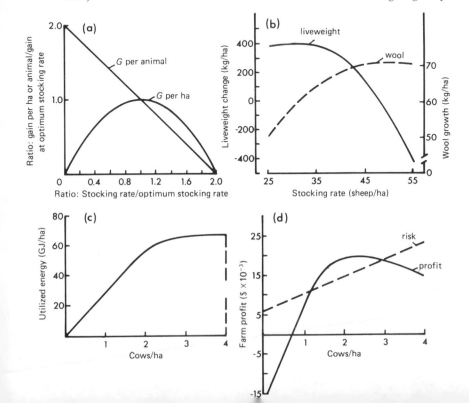

mudagrass (*Cynodon* spp.), and *b* is a measure of how liveweight gain is reduced with increasing stocking rate (Jones & Sandland, 1974).

The relationship between liveweight gain per hectare (G_H) and stocking rate is a quadratic function of the form:

$$G_H = (a \times N) - (b \times N^2) \tag{7.2}$$

The maximum gain per hectare occurs when $N = a/2b$; Jones & Sandland (1974) defined this point as the optimum stocking rate (Fig. 7.3a). But by expressing the liveweight gain per head and per ha as a ratio of the calculated weight gain at the optimum stocking rate (*y*-axis, Fig. 7.3a) and stocking rates as ratios of the optimum stocking rate (*x*-axis, Fig. 7.3a) it is possible to derive the following regressions:

$$G = 2.0 - 1.0N \tag{7.3}$$

and

$$G_H = 2N - N^2 \tag{7.4}$$

which are plotted in Fig. 7.3a.

The generalized regressions shown in Fig. 7.3a should be qualified in two ways. First, at very low grazing pressure, animal intake and production do not rise continuously with a fall in grazing pressure but rather they reach maximum values according to the potential daily intake of the animal (Fig. 6.1). Consequently, as Mott (1960) proposed, productivity may follow a declining rectangular hyperbola with respect to increasing grazing pressure:

$$G = k - ab^N \tag{7.5}$$

where *k* is an empirical constant: maximum productivity in a particular grassland. Whereas the generalized situation (Fig. 7.3a) has a broad and symmetrical optimum for production per ha, in Mott's scheme animal production per ha is skewed: it reaches a maximum beyond which production per ha decreases sharply. Rapid degeneration as the stocking rate increases beyond a 'crash point' appears to be related to a degeneration in species composition and persistence in extensive systems whereas it is not found in intensive, highly-productive grasslands where Mott's model may be more applicable. For some grass–legume pastures B. Walker (1986, personal communication) postulated the existence of a critical stocking rate below which pasture stability and a satisfactory legume content is maintained.

A second qualification upon Fig. 7.3a is that the theoretical broad optimum for stocking rate (*N*) or, more correctly, potential intake relative to the amount of available herbage (*P*), refers to the production of beef or milk per ha. The production of wool shows quite a different response than does the relationship of liveweight to stocking rate (Fig. 7.3b). Moreover, energy utilization (a measure of the efficiency of the system), like wool production, plateaus at a high stocking rate (Fig. 7.3c) while farm profits and risks have unique responses to stocking rates (Fig. 7.3d).

Grazing pressure and *P* were considered in relation to animal intake in Section 6.4. Later in this present chapter, they are considered as one variable of management which interacts with the grazing interval and rotation (see later figures). Before passing to these variables of management, however, it is worth noting again that the stocking rate directly affects the composition of grassland and thus its long-term productivity. For example, Curll & Davidson (1983) showed that high stocking rates led to dominance of annual grass (*Hordeum leporinum*) and Parsons *et al.* (1983b), Williams (1984) and Curll *et al.* (1985) found that they increased the population of ryegrass tillers, which was accentuated over successive seasons. By contrast, some legumes, e.g. lucerne (Leach, 1978) and *Desmodium intortum* (Bryan & Evans, 1973) are relatively intolerant of high grazing pressure. Botanical changes, while accompanying high levels of animal production per ha, may thus be deleterious to the long-term nitrogen economy and quality (nutritive value) of the grassland: short-term gains in productivity per ha may be partially offset by the need to renovate or resow grasslands more frequently when they are grazed at high stocking rates than when they are grazed at low stocking rates.

7.2.2
Herd experience, composition and timing of operations

Animal intake, and thus changes in milk production, depend on previous grazing experience. Sheep accustomed to high grazing pressure eat longer, and therefore have higher intake, than sheep which have grazed previously at low grazing pressure (e.g. Curll & Davidson, 1983); animals previously exposed to toxins may have greater tolerance than animals not so exposed (Section 6.2).

Herd composition depends on:
(i) the choice of livestock operation, e.g. beef fattening versus cow–calf operation; and
(ii) the timing of operations such as mating and culling.

These choices are the most obvious and easiest ones open to the farm manager.

The consequences of mixed grazing systems, e.g. cattle plus sheep compared with single species, have been studied in relation to their grazing, production and disease and pest control (Nolan & Connolly, 1977). A substantial advantage of grazing by cattle followed by sheep is to reduce worm loads (Morley &

Donald, 1980). The composition and weed content of the grassland depend on whether it is grazed by sheep or cattle, a mixed herd possibly giving better overall weed control (Watkin & Clements, 1978). Morley (1981) concludes that mixed herds are generally beneficial but that their advantages do not justify the substantial investments in fencing etc. that are necessary when shifting from one class of livestock to two. Where investments have already been made the mixed operation has small but significant biological advantages and it may be more stable financially than a single livestock operation.

Given a particular herd structure, there is still considerable scope for influencing the efficiency of the grassland system through the timing of management actions. The timing of operations determines the closeness of match between the availability of feed and herd needs. The seasonality of grassland growth, particularly in wet-and-dry climates, means that the optimum stocking rate and potential animal productivity fluctuate continuously throughout the year (Table 7.1).

Seasonal fluctuations in the optimum stocking rate may be accommodated by selling stock at the beginning of the season of feed shortage, optimizing the time of calving with respect to feed supply and sale and strategic supplementary feeding. Generally, changing the timing of operations increases calf weight at sale and improves the pregnancy rate and percentage of progeny sold (e.g. Kothmann & Smith, 1983). For example, Fig. 7.4 shows that lamb weight at weaning increases almost linearly with increases in the amount of available feed at lambing and that the

Table 7.1. *Changes in the optimum stocking rate (SR) and animal productivity throughout the year, for cattle grazing tropical grass–legume pasture on Cape York Peninsula (latitude 12°S)*

	Wet season	Dry season
Optimum SR (animal/ha)	2.1	0.42
kg gain/(head day) at optimum SR	0.65	0.13
Maximum gain/(head day)	1.35	0.054

Source: Edye, Williams & Winter (1978).

weight at weaning was almost twice as high when lambs grazed grassland growing at 35 kg per ha per day than at 16 kg per ha per day. However, many experiments, conducted in areas as diverse as the arid grasslands of Australia and the hill grasslands of Scotland, have shown that it is not economic to alter substantially the seasonality of feed supply: the most economic systems are those which accept high weight gain during the season of active grassland growth and corresponding losses over the season of poor growth (e.g. Chestnutt, 1984; Speedy, Black & Fitzsimons, 1984).

7.2.3
Grazing interval: set stocking and block grazing
Fig. 7.5 shows the pattern of milk production from cows grazed rotationally on four fields. The milk

Fig. 7.4. Production functions for lamb liveweight per ha at weaning versus stocking rate at two levels of herbage growth (a) 16 kg per ha per day and (b) 35 kg per ha per day. The gain in liveweight depends highly on the amount of herbage available at lambing, which ranged from 600 to 1200 kg per ha. (From Bircham, 1984.)

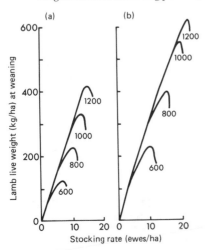

Fig. 7.5. (a) Milk production from cows under rotational grazing is associated with daily changes in (b) the digestibility of feed in the digestive tract (——) and the digestibility of the feed on offer (---). 1/8 days, 2/8 days, etc. refer to the number of the grazing and the duration (period) of grazing. (From Blaser, 1982.)

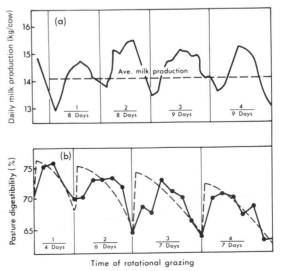

production follows closely the digestibility of feed in the animals' digestive tracts (Fig. 7.5b). On entering an ungrazed field there is a period of 1–1.5 days of low production as poor quality feed eaten in the previous field passes through the rumen; there may then be a short-term peak in milk production because the animals select better quality feed than the average for the pasture, and finally, production declines in association with declining quality and quantity of pasture. This final decline in productivity is also associated with changes in animal behaviour and eating habit (Section 6.4).

Observations such as those detailed in Fig. 7.5 have supported almost continuous experimentation since Woodman & Norman (1932) identified two substantive variables of grazing: grazing interval and (associated with it) the number of fields around which animals are grazed. It is now recognized that there are no general rules as to which interval between grazing and which degree of rotation will be optimal. Under idealized conditions (when the rate of grassland growth is constant) computer models predict that grass growth will be much the same under continuous grazing as with rotations of up to about 25 days. However, if the time elapsed between grazings of a particular field or the degree of field subdivision are further increased then the grassland becomes less productive and the amplitude of feed availability (difference between the maximum and minimum amount of feed on the field) becomes greater (Fig. 7.6).

In non-ideal, real grassland systems, the most appropriate rotation requires management of grazing to ensure productivity and persistence or regeneration of the species. Thus, for *Bromus catharticus* (prairie grass), severe, infrequent grazing results in retention of a higher percentage of prairie grass (75 per cent) than severe, frequent grazing (41 per cent) (Alexander, 1985). Furthermore, the frequency of rotation affects grassland yield and the degree of trampling, and differences in productivity, composition of grassland, etc. between rotations which are apparent after one year may become greater, or disappear, after several years (Table 7.2). Others (e.g. Wilman & Asiegbu, 1982) have found that increasing the interval between harvests (from 3–4 to 8–12 weeks) may not change the composition of white clover–ryegrass pasture but it can increase the yield; Wolton, Brockman & Shaw (1970) and Curll & Wilkins (1985) found that long intervals between grazing were advantageous for clover while Brougham (1959) and Donald (1963) showed that too lax a grazing regime may eliminate clover because of over-shading by grasses. Orr & Newton (1984), comparing grazing cycles of 18–42 days for sheep on ryegrass in England concluded that

frequent grazing of light herbage masses or infrequent grazing of heavier masses made little difference to animal performance. Thus, on a year-round basis, there is a consensus that intensive management (short grazing interval, many fields) gives benefits of only up to about 10 per cent more herbage (McMeekan & Walshe, 1963; Holmes, 1980).

Flexible rotational grazing, or block grazing as it is becoming known, may have a role when it is introduced for only part of the year. Within the context of feed year planning the most appropriate grazing strategy (set stocking or rotational grazing) depends on:

(i) the season and therefore the likely grass growth rate;

(ii) the availability of pre-grazing dry matter; and

(iii) the stocking rate.

Milligan (1984) suggests that in New Zealand rotational grazing is likely to be most effective if the first two (growth rate and availability) are low and the stocking rate is high. In practice, all three vary seasonally. Thus, in the hill country of New Zealand rota-

Fig. 7.6. Predicted patterns of herbage weight (a) and amplitude of variation in weight (b) under various grazing strategies. t, Rotation period (days); n, number of subdivisions (sub-plots). In (a) C represents continuous grazing; in (b) the continuous line represents the maximum amount of herbage accumulated at the end of the rest period and the dashed line represents the minimum amount of herbage at the end of grazing, for each of three rotations. (From Noy-Meir, 1976.)

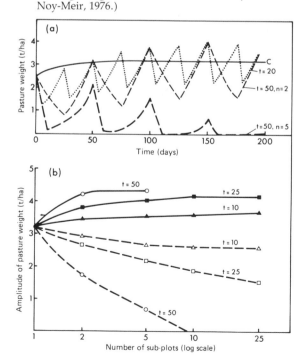

tional grazing is recommended in autumn and winter, shifting to set stocking at about the start of lambing, which should coincide with available feed reaching 1 t per ha (Bircham, 1984; Milligan, 1984). Conservation of feed is necessary in the spring to avoid deterioration in the quality of available feed and to ensure that there are sufficient supplements to maintain the intake of sheep during the winter; in this situation (north New Zealand) the requirement is for 50–100 kg dry supplement per ewe.

7.3
Conservation and supplementation

Conservation of forage involves transferring energy stored in plant material over time to meet seasonal periods of feed deficiency of grazing livestock or to feed 'housed' ruminants in non-grazing systems. Conservation includes the production of forages for sale. Forage sources include pasture, crop residues, crop and industrial by-products and grain.

It is not possible to be prescriptive about the role of conservation in farming systems. For example in New Zealand, 5–7 per cent of the area of improved grassland is conserved yet in most of southern Australia hay and silage are uneconomic. Fig. 7.7 shows the many factors which interact on the profitability of conservation. Locking up an area for conservation involves some opportunity cost; it is sometimes more desirable to manipulate the stocking rate to utilize excess feed or it may be more economic to 'lose' or not utilize excess growth. Computer simulation is probably the most appropriate tool for predicting the economics of forage conservation (Kaiser & Curll, 1986).

A range of conservation strategies exists:
(i) carryover feed which is grazed, i.e. pasture which is not utilized in the season in which it was grown but left standing in the field to be grazed in a subsequent season;
(ii) carryover of feed which is harvested manually or mechanically, stored and transported, usually on farm;
(iii) the growing of pasture or fodder specifically for conservation and sale either locally, regionally, nationally or internationally;
(iv) the sowing of crops and feeding of their residues, i.e. the 'stover' remaining after harvest of the grain;
(v) the feeding of industrial by-products; and
(vi) the sowing of a fodder crop.

Other strategies, not considered in detail here, include:
(i) the use of irrigation or fertilizer nitrogen to extend pasture growth (e.g. Murtagh, 1980);
(ii) the feeding of grain produced from associated cropping activities or purchased off-farm; and
(iii) livestock husbandry adjustments such as accepting a lowering of feed intake by grazing animals with accompanying production losses, or reducing stocking rates (Section 7.4).

Supplementation is defined here as the addition of specific components to the diet of ruminants to overcome deficiencies. The deficiencies may be highly specific, e.g. a deficiency in an amino acid, a mineral (e.g. phosphorus) or a vitamin, or they may be more general, e.g. energy or protein deficiencies. Supplements may be fed with pasture or conserved forages. In some instances pastures themselves may be seen as

Table 7.2. *Effect of grazing rotation (six-week or three-week grazing periods) and severity of defoliation on the productivity of prairie grass* (Bromus unioloides) *pasture in New Zealand*

	6-week rotation		3-week rotation	
	Lax[a]	Severe[b]	Lax[a]	Severe[b]
Year 1				
Green herbage (t/(ha y))	17.1	17.7	15.4	18.3
Trampled (t/(ha y))[c]	5.4	3.5	2.1	2.2
Year 3				
Green herbage (t/(ha y))	13.5	17.2	14.7	14.0
Trampled (t/(ha y))[c]	2.7	1.3	0.3	0.3

[a] Grazed to 7.5 cm above ground.
[b] Grazed to 2.5 cm above ground.
[c] Feed not able to be utilized (lost) due to trampling by grazing stock.
Source: Alexander (1985).

supplements, e.g. a field of legume can serve as a protein supplement in a grazing system based on poor quality native grassland.

7.3.1
Carryover feed

The simplest form of conservation is carryover pasture. Pasture accumulated during periods favourable for growth is 'saved' for periods of deficiency. Saving may involve exclusion of livestock, by fencing, or may be a self-regulating system as is the case in some tropical grass–legume pastures where animals actively select grass in the wet season and then switch to legume at the commencement of the dry season as the grass component declines in quality (Fig. 7.1).

Obviously, carryover pasture applies only in systems where year-round grazing is possible; such systems are usually found in tropical, mediterranean and mild temperate regions. The incorporation of perennial and annual legumes into the extensive grazing systems of tropical northern Australia and South America is in part a strategy to carry higher quality

standing legume into the dry season so as to maintain animal liveweight or frequently to minimize liveweight losses (Gillard, 1982). In cases where the legumes remain green through the dry season, animal performance is enhanced. High liveweight gains (over 1 kg per head per day) during the dry season have been recorded in cattle grazing the browse shrub *Leucaena leucocephala* in central Queensland, Australia. Short-lived perennials or annuals provide higher quality standing dead material than the native or naturalized grasses during the dry season when there is little to no rainfall to cause spoilage of the dead material (McCown, Wall & Harrison, 1981). Standover dead material, particularly legume leaves, and the stems and burrs of annual clovers and medics significantly contribute to animal production in the mediterranean regions of southern Australia. In these systems it is possible to select species in which standing dead material has a high nutritive value or those which retain a great amount of summer green material. In temperate regions autumn growth may be 'saved' by excluding grazing for subsequent utilization in colder winter months when there is no growth.

Fig. 7.7. Schema showing factors which interact on the profitability of conservation.

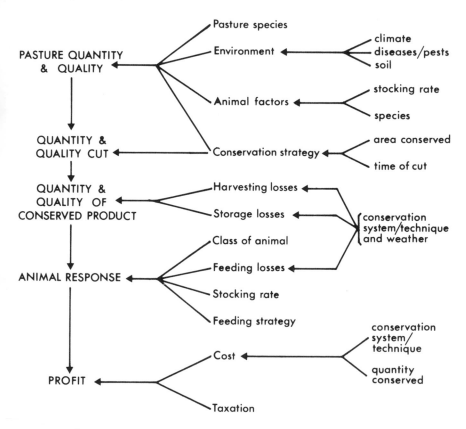

7.3.2
Hay and silage

Hay and silage are the most common forms of cut and stored pasture; they are achieved by dehydration (hay-making) and anaerobic fermentation (ensiling). The quality and hence value of hay and silage depends on:

(i) the quality of the parent material: at best conservation can only maintain quality but loss in quality is most likely;

(ii) the time of harvest in relation to the stage of plant development, amount of growth and weather conditions;

(iii) the proportion of higher quality product (e.g. leaf or legume) in the pasture; and

(iv) the extent of harvest and storage losses.

Pasture is harvested at a phase of development which gives the best compromise between maximizing the quantity and maximizing the potential quality of the conserved material. The trade-off between yield and quality indicators (digestibility and metabolizable energy) is indicated in Fig. 7.8 for S24 perennial ryegrass growing in southern England.

Green pasture continues to photosynthesize and respire for a short time after cutting; sugars then start

Fig. 7.8. Parameters involved in deciding the optimum time to harvest ryegrass grown for hay or silage in southern England. The optimum time is a compromise between yield and quality (the digestibility or D-value) and depends on the end-use. EE, date of 50 per cent ear emergence; OM, organic matter; ME, metabolizable energy. (From Raymond, Shepperson & Waltham, 1978.)

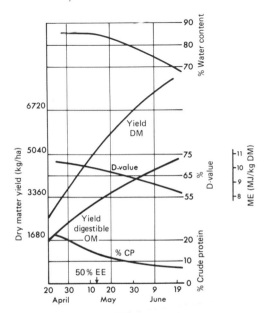

to oxidize and proteins break down and at the same time resident bacteria and fungi, innocuous on the living plant, commence attacking and decomposing dying tissues. Conservation aims rapidly to check these destructive processes; for example, respiration losses may amount to as much as 7 per cent of the original dry matter in ryegrass and white clover cut for hay (Harris & Tullberg, 1980).

Dehydration most commonly occurs from field-drying; solar radiation has a dominant effect by heating both the ambient air and the crop. Wind accelerates evaporation. Energy for dehydration is also derived from self-heating of the crop through continued respiration until the moisture content reaches 40 per cent, and from the metabolic activity of microorganisms. Dehydration is initially rapid while moisture is lost from leaf and stem stomata and the cuticle and then slows as deep tissue is dried. Leaf/stem ratios affect the drying rate due to the higher proportion of deep tissue in stems.

The rate of drying is the main determinant of field losses. The rate of drying may be increased by 'conditioning' of the hay at harvest by mechanical or chemical (e.g. potassium carbonate) means to damage the plant surface tissue and so accelerate the loss of water. Likewise, dehydration is accelerated by treatment with hay preservatives, e.g. propionic acid or its ammonium salts and anhydrous ammonia. Barn-drying, where moist hay (60–65 per cent dry matter) is put under cover at an early stage and then dried rapidly, is an expensive system developed to minimize field losses in northern European and North American grassland systems.

Poor hay-making may lose as much as 30 per cent of the total dry matter (Harris & Tullberg, 1980) or 50 per cent of the more desirable leaf fraction (Alli *et al.*, 1985) due to mechanical fragmentation. The practice and role of hay-making in farming systems is further reviewed by Klinner & Shepperson (1975), Jarrige, Demarquilly & Dulphy (1982), Kaiser & Curll (1986) and Bolsen (1986).

The principles of ensiling, the process of producing silage, were established by the early part of this century (McDonald, 1981; Jarrige *et al.*, 1982). The principles involve exclusion of air (anaerobiosis) so that bacteria which produce lactic acid can dominate the fermentation. These ever-present lactobacilli grow on a substrate of water-soluble carbohydrate. The growth of undesirable clostridial bacteria, moulds and fungi are suppressed. This depends on a favourable fermentation which reduces rapidly the pH of the silage. The clostridial species of bacteria are unable to survive in acid conditions (usually less than pH 5) although there is an interaction with moisture content:

at lower moisture contents a higher pH can be tolerated (Ross, 1984).

Good silage requires that the pasture initially has a high dry matter content of 300 g per kg or more. This relates to the need to acidify the silage rapidly: the percentage of dry matter determines the critical pH value which prevents clostridial development. The dry matter content is increased by wilting prior to ensiling.

Ensiling involves physical damage (bruising, laceration, mincing and chopping) of forage to release soluble carbohydrates, to improve consolidation in the silo and to increase its acceptability to animals. Silos vary in size and shape and in the materials from which they are constructed. They range from high cost metal structures to relatively simple earthen pits, although increasingly plastic is being used to create anaerobic conditions.

It is desirable that minimal losses of nutrients occur if silage is to be useful for animal production; as in hay-making there is a range of processes and stages during which losses occur (Table 7.3). Losses are higher in tropical than in temperate farming systems (Wilkinson, 1983a, b) to the extent that there is little role for hay or silage in the tropics. Jarrige *et al.* (1982) consider the role of hay and silage in the tropics in more detail.

7.3.3
Crop residues and by-products

Crop residues refer to the mature plant material left standing on the ground or accumulated after removal of the crop grain, tuber or fruit. Residues include dead stems, leaves and husks. In some instances, livestock will also have access to spilt grain and weeds growing in the stubble. By-products are produced during the processing of crops or agro-industrial commodities which are unsuited to human or monogastric (pigs, poultry) consumption or for which there is no demand. The terms residue and by-product are not mutually exclusive. In addition crop material from thinning, topping and leaf-stripping may be fed to animals.

The usefulness of residues and by-products depends on their initial quality and the technology available to improve their quality, the animal productivity obtained from their use and the economic advantages these materials have over the other feed sources in particular farming systems. They may be used to complement the use of hay, silage and pasture or they may provide the basis of seasonal or year-round feed systems. Depending on their quality and mineral balance, they may also be used as supplements.

Non-livestock uses of residues include industry (for fuel, paper, alcohol and fibre board) and in conservation farming (for erosion control, soil amelioration and water conservation). Thus there may be conflicting demands for residues. For example, straw may be used for livestock bedding and feed or it can be used in soil conservation.

The role of crop residues in livestock feed systems is reviewed by Jarrige *et al.* (1982), Wilkins (1982), Ibrahim (1983), Mulholland, Coombe & Pearce (1984) and Dann & Coombe (1987). The value of residues to livestock is limited by the poor quality of residues and their consequent low utilization. Crop residues and

Table 7.3. *Typical losses of dry matter during the conservation of crops as silage in bunker silos under tropical or temperate conditions*

Loss (%)	Tropical	Temperate	
	Direct-cut	Direct-cut	Wilted
In field			
Respiration	—	—	2
Mechanical	6	1	4
During storage			
Respiration	—	1	1
Fermentation	7	5	5
Effluent	—	6	—
Surface waste	7	4	6
During removal from store	5	3	3
Total	25	20	21

Source: Wilkinson (1983a).

associated herbage are usually deficient in nitrogen and some minerals and their digestibility is usually less than 50 per cent (Fig. 6.5).

Animals grazing crop stubble usually maintain their liveweight or even gain weight in the first two weeks. Thereafter livestock lose weight.

Utilization of stubbles by grazing animals is low. Mulholland *et al.* (1984) cite maximum utilization levels of available dead material as 36 per cent, equivalent to 25 per cent of the total straw available at the beginning of grazing. Fifty per cent of available straw disappeared via other pathways including decomposition, wind transport and treading by stock.

Physical, chemical and biological pre-treatments are technologically feasible for improving the quality of residues but most are energy intensive and uneconomic. Examples of physical pre-treatment are grinding, ball milling, soaking, boiling, steaming, irradiation, explosion and pulping. These treatments reduce particle size, which increases voluntary intake and the efficiency of utilization in the rumen due to the increased rate of passage of ingesta comprised of small particles from the rumen (Wilkins, 1982). The energy which can be utilized through digestion by the animal is also increased in materials which have low digestibility: chemical pre-treatments with sodium hydroxide, ammonium hydroxide, calcium hydroxide, sodium chlorite, sulphur dioxide, chlorine, ozone and sulphuric acid aim to improve the accessibility of structural carbohydrates to rumen microbial breakdown; success has been greater with lower quality grass or cereal residues than high quality feeds including

legume pasture and stubble. Physico-chemical treatments usually involve hydroxides (alkalis) combined with steaming under pressure or explosion pulping. Biological and biochemical treatments, such as white rot fungi and enzymes aim to increase the digestibility of the lignocelluloses. This is an area of as yet unrealized potential (Wilkins, 1982).

Management, selection and breeding may affect the quality and utility of residues. The widespread adoption of short-strawed high-yielding rice and wheat varieties in Asia has had detrimental effects on livestock where straw is a major component of the diet. In India straw to grain ratios of modern and local cereal varieties average 1.7 and 2.2 respectively. There has been a consistent decline in the digestibility of wheat and rice straw from 45 to less than 40 per cent. This drops the digestible energy concentration of the crop residue from a maintenance to a sub-maintenance diet (Deboer, 1983).

By-products are usually localized, seasonal, of variable chemical composition, limited in transportability owing to high moisture content or low bulk density and possibly contaminated with e.g. heavy metals, bacteria and toxins or anti-nutritional factors (Low, 1984). In most farming systems there is the potential to utilize by-products when either the technology is developed or the economic conditions are suitable (e.g. Ranjhan & Chadhoker, 1984 for Sri Lanka). Distance from the source and thus transport costs is frequently a key economic criterion for the utilization of by-products. The characteristics of different categories of by-product are summarized in Table 7.4.

Table 7.4. *Characteristics of categories of by-product used to meet seasonal deficits in feed*

	Digestibility	Nitrogen	ME[a]	Examples	Comments
(i)	Low	High	Low	Animal wastes (cattle, poultry)	Cannot be used as sole feed; must combine with high energy feeds, e.g. grain
(ii)	High	High	Variable, usually high	Whole cottonseed; cottonseed, soybean, sunflower meal	Can be used at high levels; cost and sometimes anti-nutritional factors may limit use
(iii)	High	Low	High	Citrus pulp, molasses	Must be incorporated with feeds with adequate nitrogen, e.g. (i)
(iv)	Low	Low	Low	Straws, hulls, husks, wood by-products	Can be used as diluents or fillers but not adequate alone; treatment may improve quality (Chapter 8)

[a] Metabolizable energy.
Source: Adapted from Low (1984).

7.3.4

Fodder crops

Crops may be sown specifically for complete utilization, or for partial utilization by livestock on a planned or opportunistic basis. They may be grazed, mechanically harvested or both. Fodder crops may be summer- or winter-growing species (Table 7.5). Dann & Coombe (1986) express the rationale for the sowing of fodder crops as the expectation that they will provide more and possibly better quality feed than will permanent grassland at that time of the year when crops are grazed or harvested.

The utility of fodder cropping depends on:

(i) the relative animal production from the sown crop and the grassland it replaces;
(ii) the direct costs (e.g. of seed, fertilizer) of sowing the crop;
(iii) the opportunity cost of the crop, i.e. the length of time land is taken out of production for preparations for sowing and during establishment and the concomitant increases in grazing pressure on the rest of the farm; and
(iv) the availability of stock to utilize the crop.

Wheeler (1981*a, b*) reviews this subject. It is difficult to be prescriptive about the potential role of fodder crops in farming systems. Their contribution to animal production and thus economic worth are difficult to measure and Dann & Coombe (1987) suggest that computer modelling may be the only effective means of assessing the role of fodder crops in whole farm systems.

7.3.5

Supplements

Supplements may supply deficient nutrients, substitute for part of the diet or increase forage consumption.

Deficiencies in diets are usually of energy, protein, minerals and vitamins. Two situations of energy deficiency are commonly experienced in grazing systems:

(i) Gross energy deficiences due to drought, flood or fire when supplementary feeding is needed for survival of grazing animals. In such circumstances the most readily available cost-effective feed source is used or alternatively animals may be sold. Grain is usually the most cost-effective supplement; in the semi-humid tropics stored crop by-products and residues such as rice straw, rice bran, cassava meal, leguminous browse shrub

Table 7.5. *Some commonly-sown winter-growing and summer-growing fodder crop species*

Summer		Winter	
Grasses			
Japanese millet	*Echinochloa utilis*	Oats	*Avena sativa*
Sorghums	*Sorghum* spp.		*Avena byzantina*
Sudan grass	*S. sudanense*		*Avena strigosa*
Sorghum × Sudan grass hybrids		Wheat	*Triticum aestivum*
Maize	*Zea mays*	Barley	*Hordeum* spp.
		Triticale	*X. Triticosecale* spp.
Pearl millet	*Pennisetum americanum*	Rye	*Secale cereale*
Millet hybrids		Tetraploid ryegrass	*Lolium* spp.
Legumes			
Cowpea	*Vigna unguiculata*	Persian clover	*Trifolium resupinatum*
Lablab	*Lablab purpureus*	Berseem clover	*T. alexandrinum*
Sunnhemp	*Crotalaria juncea*	Arrowleaf clover	*T. vesiculosum*
Mung beans	*Vigna radiata*	Lupins	*Lupinus angustifolius*
Pigeon peas	*Cajanus cajan*		
Brassicas		Rape	*Brassica napus* var. *napus*
		Kale	*B. oleracea* var. *acephala*
			B. rapa var. *rapa*
		Swedes	*B. napus* var. *napobrassica*
		Fodder beet	*Beta vulgaris*

and lopping of large non-leguminous trees may be the only alternatives (e.g. Nitis, 1986).

(ii) Energy deficiencies due to seasonal drought or cold and where feeding for production is feasible. Here the aim is to maximize the intake of available forage (Siebert & Hunter, 1982). Supplements of readily available energy will stimulate organic matter digestion in the rumen and increase the feed intake of low quality roughages if these supplements are fed continuously. However, if the supplement is fed intermittently, i.e. not continuously, then there is substitution of the high for the low quality diet and changes in the rumen microflora (Liebholz & Kellaway, 1984).

Protein deficiency occurs when the nitrogen requirement of the animal is not met from its diet. The nitrogen requirement of ruminants is met from three sources:

(i) Rumen-degradable nitrogen (RDN), which is used for microbial synthesis of amino acids. This may originate as soluble dietary protein from fresh pasture or as non-protein nitrogen, e.g. urea.

(ii) Undegraded dietary protein (UDP) or by-pass protein which escapes fermentation in the rumen and passes to the small intestines. Tables of UDP and RDN requirements for cattle and sheep at various levels of production are published by the Agricultural Research Council (ARC) (1980).

(iii) Endogenous amino acids from secretions in the stomach and sloughing of epithelial cells (Liebholz & Kellaway, 1984).

Where available pastures or residues are low in protein or nitrogen and sulphur concentration, but sufficiently digestible to provide energy to rumen microbes, then supplementation with RDN protein increases the net intake of forage by increasing the rate of digestion and the rate of passage of digesta (Siebert & Hunter, 1982). Protein supplements include various crop meals such as peanut, cotton and linseed meals, and meals from other sources such as meat and fish meal. Standing, green or dead legume herbage can be seen as providing a protein supplement to poor quality grass residues in the seasonally-dry tropics and in summer-dry mediterranean regions (e.g. Allden, 1982; Winks, 1984).

Mineral supplements include sulphur, calcium, phosphorus, sodium and magnesium and micro-nutrients of which only cobalt, copper, iodine, selenium, manganese, molybdenum and zinc have been recorded as giving a response in animal production (Siebert & Hunter, 1982). Mineral deficiencies may be clinical but are usually sub-clinical, not readily identified and capable of reducing productivity by

say, 10 per cent (Judson et al., 1986). Mineral supplements may be specific, e.g. cobalt pellet or bullet given orally to the animal, or iodine given intravenously. They may also be broad spectrum, e.g. superphosphate fertilizer which contains sulphur, calcium, phosphorus and possibly molybdenum and zinc. They may have two modes of action: a primary effect on the animal and a secondary effect on pasture growth, sward composition and nutritive value.

Vitamin A deficiencies occur in livestock deprived of green feed for long periods, e.g. in droughts. Vitamin B_{12} deficiencies may result from deficiency of cobalt, one of the precursors of the vitamin, in the diet. Vitamin injections may overcome deficiencies (Judson et al., 1987).

Diet substitution is a major problem with any form of supplement: the supplement may substitute for part of the diet formerly grazed. This occurs mainly when high quality, readily-accessible supplements are substituted for poor quality forage, e.g. when feeding hay, silage, concentrates and fodder crops. Substitution may be caused by reductions in intake of the original diet to allow intake of the supplement, or because of reductions in the incentive to graze.

7.4
'Feed year' planning

Feed year planning allows the farmer or grassland agronomist to match the supply of feed with animal demand. The management objective of this is to determine the potential livestock carrying capacity of a field or farm and make appropriate adjustments in the husbandry of livestock, grassland and crops. Feed year planning may be carried out on a whole farm basis, for specific livestock units, or for particular forage sub-systems.

Planning may be conducted at various levels of sophistication: in its most sophisticated form feed may be budgeted when the 'income' is feed supply, the 'expenditure' is animal demand and the outcome reveals a surplus or deficit. Budgeting feed supply for a given animal demand is conceptually appealing; in practice its application varies with farmer circumstance and particular animal–grassland system. It has maximum utility in systems where there is little climatic variability and no, or regular, changes in the botanical composition of the grassland. This is because the first step in preparing a feed budget is to estimate the total grassland dry matter (DM) available. To do this it is necessary to (i) assess the total amount of feed present (kg DM), (ii) know the species present and predict likely changes in botanical composition, (iii) know how much of that feed can be utilized by stock for a specific productive purpose,

(iv) estimate the expected grassland growth and (v) know the quantity and quality of available supplements.

Feed supply data can be generated by five means:

(i) From farmer or advisor experience.

(ii) From paddock measurement either by direct sampling (cutting and drying), or indirectly using capacitance meters, rising plate meters or visual appraisal (see Frame, 1981). Usually some direct sampling with quadrats is inescapable if only for calibration of meters or the eye or validation of models.

(iii) From the use of production curves generated by research and found in the literature. This involves extrapolation from growth curves averaged over a number of years for specific (usually few) sites.

(iv) From empirical regression models which commonly feature the regression of the increment in dry weight versus light and temperature, a water balance model or estimate, and discounting of quality with pasture age.

(v) From mechanistic models using computer simulation of the dynamics of grazing systems.

Regression models predict feed supply by estimating pasture growth rates from dimensionless indices (Fitzpatrick & Nix, 1970) which can be multiplied by potential pasture growth rates. One simple model describing the effect of environment by indices is:

$$GI = MI \times TI \times LI \qquad (7.6)$$

where TI is the temperature index (Fig. 3.10), LI is the light index, GI is the growth index and MI is the moisture index, which may be set at 1 (optimum) when the grassland is fully irrigated. Under dryland conditions MI is estimated by a water balance budget which estimates the availability of water in the soil relative to the amount of water which is available to the plant when the soil is at field capacity (Appendix A; also Fig. 3.9).

Potential growth rates for species can be found in field data. For example, for white clover and paspalum growing in south-eastern Australia, Crofts, Geddes & Carter (1963) measured maxima of 101 and 125 kg per ha per day respectively in irrigated pastures, i.e. where moisture was considered optimal. As light and temperature were not optimal when these rates were recorded in the field, potential growth rates may be calculated as follows:

White clover: Maximum growth rate = 101 kg per ha per day when TI = 0.825 and LI = 0.91, therefore GI = 0.75
Potential growth rate = 101/0.75 × 7 = 943 kg per ha per week

Paspalum: maximum growth rate = 125 kg per ha per day when TI = 0.785 and LI = 0.93, therefore GI = 0.73
Potential growth rate = 125/0.73 × 7 = 1195 kg per ha per week

Alternatively, potential growth rates may be found from regressions fitted directly to experimental data collected in controlled temperature studies provided that the experiments are designed so that plant performance will be relevant to the field, e.g. at the same plant spacing and nutrition (Muldoon, 1979). Once potential growth rates have been derived they may be multiplied by GI calculated on a weekly basis for a given locality where meteorological records are available, to predict the actual rate of pasture growth, G:

$$G = \text{Potential growth rate} \times GI \qquad (7.7)$$

There are numerous computer-based models for predicting the availability of soil water from soil and climatic data. By contrast, few regression-based pasture budgets make provision for the effects of mineral nutrition on growth (Fig. 5.7), or of predation (Fig. 3.11) or of the environment, e.g. excess moisture or daylength, on plant development.

Mechanistic models using computer simulation are the most sophisticated method of generating estimates of food supply. The place of mechanistic models vis-a-vis regression models is discussed by Thornley (1976) and in Section 8.5. Here it is sufficient to note that they permit estimation of the importance of biological processes within the pasture production system but, because of their complexity, they are currently not as widely used as regression models for predicting the availability of pasture.

The feed requirements of grazing livestock were discussed briefly in Section 6.5. Oddy (1983) and MAFF (1975) explain the necessary calculation in detail. An example is provided in Appendix B. Various units of measurement have been employed to express both animal demand and its relationship to grassland productivity, carrying capacity or grazing pressure (Alderman, 1983). Pasture growth is commonly expressed as kg dry matter or organic matter per ha but to equate it with animal energy demand the now-accepted convention is to convert this to an MD value where this is the concentration of metabolizable energy (ME) per kg of pasture dry matter (Section 6.5). The output of feed year planning is of the form shown in Fig. 7.9.

Christian (1987) recognizes four broad strategies within feed year planning. They are:

(i) modification of animal requirements, by changing the stocking rate, herd or flock composition, date of joining, time of weaning;

(ii) optimization of grazing management, by changing the intensity of defoliation, the degree of subdivision, or choosing to defer grazing or conserve forage;

(iii) modification of grassland production, by species improvement, extending the growing season, the use of legumes, fertilizers and irrigation; and

(iv) augmentation of feed supplies, by using fodder crops, low quality residues, supplements, by-products and drought feeding.

Biological aspects of these management options were discussed in Sections 7.2 and 7.3.

The extent to which farmers manage or manipulate their feed year (e.g. Fig. 7.9b) depends on the biological efficiency within a particular technology, e.g. hay-making, and on environmental, economic and cultural constraints (Section 8.2).

7.5
Efficiency of livestock production

7.5.1
Energy budgeting

An energy budget is a summary of the 'energy operating costs' and 'outputs' of the system in units of Mega Joules (MJ). Energy budgeting requires assumptions about yields, fertilizer application rates, etc. based on farm measurements or experimental data. Energy expenditure on processes and concentration in the products of grassland systems may be calculated from Leach (1976), Stanhill (1984) and Pimentel (1980).

Energy budgets of pastoral systems have been based on the ratio of the digestible energy in the harvested forage to the input of total support energy

or on the mass of commodity produced relative to the total energy inputs in production (Heichel, 1985). Spedding (1984*a*) emphasizes that ratios are neither right nor wrong, but relevant, appropriate and useful or not, and that the time period over which the calculation is made must be specified. Energy analysis and budgeting should be seen as complementing economic analysis and budgeting; it is difficult to foresee it being used in farm management but it should contribute to sectoral and policy decision-making.

There are differences in the efficiency of use of solar energy and subsequent animal production between grassland systems based on temperate C_3-species and those based on tropical or subtropical C_4-species. Tropical grasses have twice the potential yield of temperate, C_3-species under non-limiting conditions (Section 3.5) but this potential is rarely realized in the conversion of dry matter to animal product (Wilson, 1986). This is illustrated by a comparison between a temperate *Lolium–Dactylis* grassland and a *Paspalum notatum* subtropical grassland in Japan (Fig. 7.10). The initial conversion of solar energy to herbage, the net primary productivity (NPP) (Section 3.6), is greater in the paspalum grassland; losses occur due to a variable efficiency of consumption and a lower herbage intake and feed conversion efficiency with the consequence that the energy stored in cattle is lower. The efficiency of consumption, the percentage of NPP which is eaten by all herbivores including domestic livestock, varies according to the time frame and management system; on an annual basis it is usually 10–50 per cent but it may be as high as 70 per cent in very controlled systems (e.g. Appendix 1 in Milligan & McConnell, 1976). It is less than 100 per cent because

Fig. 7.9. Seasonality of production and the output of a feed year plan. (a) The seasonal production of metabolizable energy (ME) from grass and ME requirements at three different stocking rates are shown for southern England; (b) to maintain a stocking rate of 3 cows/ha, excess herbage is conserved and fed back with 60 per cent efficiency (black) and supplemented with off-farm grain or hay (stippled). (From: (a) Doyle, Corrall, le Du & Thomas, 1982.)

Fig. 7.10. Efficiency of utilization of solar energy (per cent) for animal production: a comparison between temperate (left-hand side) and subtropical (right-hand side) grasslands. (From Okubo, Kirakawa, Okajima & Kayama, 1983.)

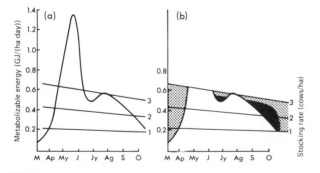

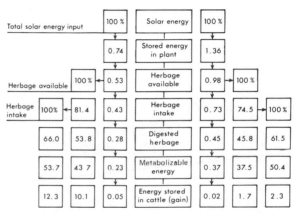

of the mismatch between the seasonality of pasture growth and the relative stability of animal requirements and because the lower parts of plants are inaccessible to the grazing animal and die before they can be consumed. It is low in rangeland ecosystems; the proportion of NPP which is consumed by large herbivores is also low, e.g. 2.9 per cent of aboveground NPP was eaten by cattle and antelope in a New Mexico grassland (Pieper, 1983). The reasons for low animal intake and production on tropical grasses were discussed in Chapter 6. Here, it is worth repeating the importance of the level of an animal's productivity to variation in gross efficiency of animal production: Table 7.6 shows that as the daily intake of energy by an animal increases, so does its production of milk, beef, wool or hide. Thus, the proportion of intake that is used to maintain the animal decreases: the gross efficiency of animal production increases as intake increases (Table 7.6).

Maximum feed intake (2100 MJ per day) cannot be obtained from grassland as its bulk is too high per unit of metabolizable energy. This high level of intake can be obtained only from grain or concentrate. However, the use of grains and concentrates for animal feed decreases the return on human edible energy and protein inputs, i.e. the use of these products to feed animals reduces the contribution of ruminant production to the total human food supply (Baldwin, 1984). The advantage and principal role of ruminant livestock is that they contribute to the efficiency of total food production because of their ability to use material which is not suitable for direct human consumption.

Much of the productivity of grasslands is in the form of animal protein. The efficiency with which

Table 7.6. *Effect of productivity as determined by feed intake on gross efficiency of a 500 kg lactating cow*[a]

Feed intake (MJ/day)	Milk energy (MJ/day)	Gross efficiency (%)
66	8	12
79	16	21
106	32	30
158	63	40
211	95	45

[a] Assumes no change in body energy content and that all animals are fed the same diet at all levels of intake.
Source: Adapted from Baldwin (1984).

protein is produced ranges from 3 to 30 per cent in California (Table 7.7). This can be compared with milk and beef production efficiencies from individual animals of 50 and 20 per cent respectively.

For domestic animals, Spedding (1971) indicates that the most useful expression of feed conversion efficiency, the efficiency of conversion of food eaten to animal product, is for one reproductive unit, i.e. taking account of the structure of the animal population. For example, considering the efficiency of meat production (E_m):

$$E_m = P/F \tag{7.8}$$

where P is the animal product (carcass weight C times number of progeny N) and F is the sum of food consumed by the dam during pregnancy (D_p), lacta-

Table 7.7. *The efficiency of animal production based on Californian production systems*[a]

	Efficiency of animal product output in relation to total feed energy (E) and protein (P) inputs		Efficiency of animal production in relation to input of human edible energy (E) and protein (e.g. grain) (P)	
	E (%)	P (%)	E (%)	P (%)
Milk	23.1	28.8	101.1	181.4
Beef: calves to feed lot at weaning	5.2	5.3	57.1	108.8
Beef: calves graze for one year before feed lot finishing	3.1	3.5	85.0	120.3
Pigs	23.2	37.8	58.0	86.0

[a] Values over 100% indicate that animals add to the total human food supply, whereas those below 100% indicate the price paid in lost efficiency for (i) food of improved nutritional value or (ii) the supply of special or desired animal products, as compared with the nutritional value to humans of the feed (usually grains) fed to produce these products.
Source: Adapted from Baldwin (1984).

tion (D_1) and the non-pregnant non lactating period (D) and the food required by each progeny (Y):

$$E_m = (N \times C)/(D_p + D1 + D(N \times Y)) \qquad (7.9)$$

Many of the feed inputs for ruminant production may be in a form already suitable for human consumption, e.g. cereals. Table 7.7 shows that only milk returns the same amount of energy as is already in the ration. In other words the pig and two beef systems in this example result in energy losses when grain is used as a feed input. In terms of energy, protein, the use of support energy and solar radiation, animal production is much less efficient than cropping (Spedding, 1984a). Spedding (1981) gives an example of sheep meat production in southern England where the energetic efficiency of gross energy in boneless carcass (i.e. meat plus fat) per unit of usable incident solar radiation is 0.02 per cent. These examples emphasize the fact that animal products which are harvested always represent only a small proportion of the nutrients and energy in the system.

7.5.2
Legume- versus nitrogen-fertilized grass systems

For both cattle and sheep the rates of liveweight gain, herbage intake and efficiency of feed utilization are greater on grass–legume mixtures than on grass alone although the rate of animal production per ha may be lower from the legume (Reid & Jung, 1982). The ratio energy output to energy input of grass–legume pastures may be five times greater than that for grass plus nitrogen (Gordon, 1980: Table 5.4). Lack of stability and difficulties of management of legume pastures are reasons frequently cited for the predominance of nitrogen-fertilized grasslands in much of northern Europe and America. The utility of each system in North America (Burns & Standaert, 1985), New Zealand (Ball & Field, 1985) and Australia (Myers & Henzell, 1985) has recently been reviewed. The cost of fertilizer nitrogen is obviously a key determinant; in the period 1970–80, nitrogen prices tripled in America without commensurate increases in the prices of animal products (Reid & Jung, 1982). Jacobs & Stricker (1976) and Spedding (1981) suggest increasing the emphasis on legume–grass systems because of their animal production, energy and economic efficiencies.

7.5.3
Efficiency of use of support energy

Support energy is used in intensive grassland production, harvesting and storage as well as in animal management and 'downstream' or off-farm activities such as transport, packaging, manufacture and distribution.

The energy in animal products per unit of energy input is low: between three and ten times as much support energy is used in production as is actually produced (Leach, 1976). While the ratio of energy output to input for the total energy production from grass or legume is relatively high, the conversion of this material to, say, fuel or extracted protein (by fractionation) (Section 8.1) reduces the efficiency of grassland below that of crops. In farms there is a range of efficiencies associated with pasture and animal production: in Japan the ratio of energy output to input for a number of pasture and fodder systems ranges from 2.5–1 for oats plus maize silage to 8.6–1 for grazed *Paspalum notatum* and wild *Vicia* (Okubo, 1982).

Grazing systems are more energy efficient than conservation or fodder crop systems, which have high inputs of support energy. However, grazing and browsing do have energy costs (Section 6.5.3). A comparison of the efficiency of use of support energy of six 'harvesting' strategies in the United Kingdom revealed (White, Wilkinson & Wilkins, 1983) that:

(i) in most instances fertilizers were the main source of energy input, in terms of support energy per unit of metabolizable energy output, although energy used in drying was the greatest for barn-dried hay;

(ii) in nitrogen-fertilized grass systems grazing did not offer as great an energy saving as expected when compared with grass which was grown for longer periods and mechanically harvested due to the need for more frequent applications of fertilizer under grazing (on ryegrass–white clover pastures grazing was three and seven times more efficient than zero grazing or wilted silage respectively);

(iii) legume-based pastures were more energy efficient than pastures based on perennial ryegrass and nitrogen fertilizer;

(iv) the energy output per unit area of land was greatest for grazing and least for field-cured hay, but greater for nitrogen-fertilized ryegrass than for lucerne and ryegrass–white clover mixtures; and

(v) feed was the major energy input in all feed conservation systems.

7.6
Further reading
Holmes, W. (ed.) (1980). *Grass, its Production and Utilization.* Oxford: Blackwell. Three chapters on grazing management, conservation and the economics of utilization.

Nestel, B. (ed.) (1984). *Development of Animal Production Systems.* Amsterdam: Elsevier. Regional livestock production; general chapters such as that by Spedding include e.g. efficiency.

8

Grasslands in farming systems

In this chapter we examine the role of grassland agronomy in farming systems from a holistic or systems perspective. The grassland or pastoral system was modelled in Chapter 1 (Fig. 1.2) and the following chapters expanded upon parts of the model and its dynamic nature. Implicit in our model is the knowledge that the grassland system is but one sub-system of a farming system (Fig. 1.1). It is our thesis that grassland agronomy and the grassland system ultimately cannot be considered in isolation either from other components of the farming system or from externalities such as society's cultural (political and social) values, which impinge on farming (Fig. 8.1).

8.1
Farming systems perspective and terminology

A farming system may be seen as an arrangement of activities which a farmer or farm household manages according to goals and in response to environmental, evolutionary, economic, technological and cultural forces. The boundary of a farming system is defined by what the manager is able to control. The farming system is part of larger systems such as agroecosystems and local communities; it can be divided into sub-systems such as the grassland or pastoral system and cropping system (Fig. 8.1).

A farming system shares the same concepts, properties and terminology as a grassland system. These are listed in Table 8.1. The purpose and measures of performance of a farming system are predominantly social and economic in nature: farming systems exist to fulfil a variety of human-set goals (Clawson, 1969; Spedding, 1984b). By contrast, for the grassland sub-system there is, strictly speaking, no purpose and the output is biological or the direct monetary value of biological products: biological components interact (grow, eat, etc.) according to biologi-

cal principles within the sub-system and the grassland sub-system is managed or regulated by man (Fig. 8.1).

Two pairs of systems properties are particularly relevant to the relationship between the farming system and the grassland system. These are hierarchy and emergent properties, and control and communication. We have already recognized a hierarchy proceeding from a plant to a grassland system to a farming system (Fig. 1.1). An understanding of a lower level in the hierarchy, e.g. a grassland, does not necessarily lead to an understanding of a farming system: the whole is different from the sum of the parts, and only when the whole is studied do all its properties emerge.

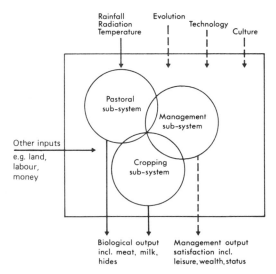

Fig. 8.1. A model of a farming system as a human activity. Sub-systems of major relevance are identified, as are major inputs and outputs which are quantifiable (⟶) or not readily quantifiable (--→).

Control may be defined by the boundary of a system: the boundary encloses those sub-systems over which the system has a reasonable degree of control. This control is achieved by communication; by managing the system through exchanges of information between the sub-systems and also through exchanges between the system and the wider systems within which it operates.

8.2
Farming as a human activity

The variety and complexity of farming systems which include livestock may be gauged from details of those in Africa and Latin America (Table 8.2). All farming systems are involved with the interaction between natural ecosystems and man's social system. The subject of this book relates mainly to the former, particularly the biology of managed ecosystems. Nonetheless, biological or economic descriptions alone are not capable of fully describing:

(i) the role (value, attitude, knowledge and skill) of the farmer who is the key to management decision-making;

(ii) present or potential external forces likely to influence the role of grasslands in farming systems (Fig. 8.1); and

(iii) the problems associated with organizational change or development such as the introduction, evaluation, release and commercialization of new cultivars and management 'packages'.

The role of the manager is often, but not always, concerned with maximizing productivity or efficiency. Achievement of this purpose can be viewed as a special form of output of the management sub-system which is concerned with innovations (e.g. new technology), allocations (e.g. of land, labour) and the carrying out of the operations (e.g. sowing, fertilizing).

Having provided a farming systems perspective and identified farming as a human activity system conducted for some purpose, let us now consider the forces which constrain or enhance the management of grasslands in farming systems. The first is the way in which grassland research is conducted; research is both a property of the system and an external force upon it. Other forces which are external to the management of grasslands may be grouped into environment (natural perturbations), evolution, economics, technology, and culture (Fig. 8.1).

Table 8.1. *Generalized systems concepts applicable to human activity systems*

Concept	Definition
Boundary	The borders of the system which define where control action can be taken: a particular area of responsibility to achieve system purposes
Resources	Elements which are available within the system boundary
Transformation	Changes, modelled as an interconnected set of activities, which convert an input which may leave the system (a 'product') or become an input to another transformation
Connectivity	Logical dependence between elements (including sub-systems) within a system
Purpose	Objective, goal or mission; the 'raison d'être' which in terms of the model is to achieve the particular transformation that has been defined
Measure of performance	A measure of how well a system achieves a particular purpose
Decision-taking	Information collected according to measures of performance is used to modify the interactions within the system
Monitoring and control	Information collected and decisions taken are monitored and controlled and action is taken through some avenue of management (see e.g. Section 7.2)
Hierarchy	The location of a particular system within the continuum of biological organization (Fig. 1.1). This means that any system is at the same time a sub-system of some wider system and is itself a wider system to its sub-systems
Emergent properties	Properties which are revealed at a particular level of organization and which are not possessed by constituent sub-systems. Thus these properties emerge from an assembly of sub-systems

Source: Adapted from Wilson (1984).

Table 8.2. *Major agricultural systems with grassland components in Africa and Latin America*

Farming system	Major crops	Major animals	Main regions[a]	Feed source
Africa Pastoral herding (animals very important)	Vegetables (compound)[a,b]	Cattle, goats, sheep	Savanna (Southern Guinea)	Natural rangelands, tree forage
	Millet, vegetables	Cattle, goats, sheep	Savanna (Northern Guinea and Sahel)	Natural rangelands, tree forage, crop residue
Bush fallow (shifting cultivation, animals not important)	Rice/yams/plantains, maize, cassava, vegetables, tree crops, cocoyams	Goats, sheep	Humid tropics	Fallow, crop residues
	Sorghum/millet, maize, sesame, soybeans, cassava, sugarcane, tree crops, cowpeas, vegetables, yams	Cattle, goats, sheep, poultry, horses	Transition forest/ savanna. (Southern Guinea, Northern Guinea and Sahel)	Fallow, straws, stover, vines, roots, sesame cake
Rudimentary sedentary agriculture (shifting cultivation, animals important)	Rice/yams/plantains, maize, cassava, vegetables, tree crops, cocoyams	Goats, sheep, poultry, swine	Humid tropics	Rice, bran, roots, crop residue, vines, stover
	Sorghum/millet, maize, sesame, cotton, sugarcane, tree crops, cowpeas, yams, tobacco, groundnuts, vegetables	Cattle, goats, sheep, poultry	Transition forest/ savanna/(Guinea and Sahel)	Stover, vines, sugarcane tops, cull roots or tubers, tree forage, groundnut cake, brans
Compound farming and intensive subsistence agriculture (shifting cultivation, animals important)	Rice/yams/plantains, maize, cassava, vegetables, tree crops, cocoyams	Goats, sheep, swine, poultry	Humid tropics	Rice straw, rice bran, vegetable waste, fallow, vines, cull tubers or roots, stover, tree crop by-products, palm oil cake
	Vegetables, sugarcane, tobacco, sesame, maize, tree crops, groundnuts	Goats, sheep, poultry, swine	Transition forest/ savanna	Vines, stover, treecrop by-products, groundnut cake
	Vegetables/millet, cassava, cowpeas, tobacco, cotton, groundnuts, tree crops[b]	Cattle	Savanna (Guinea and Sahel)	Vines, tree crop by-products, cassava leaves, fallow
Highland agriculture (animals important)	Rice/yams/plantains, maize, cassava, vegetables, cocoyams	Goats, sheep, poultry, swine	Humid tropics	Fallow, leaves, stover, rice by-products, cull tubers, cassava leaves, vegetable residues
	Sorghum, soybeans, cowpeas, cassava, maize, millet, groundnuts	Cattle, goats, sheep, poultry	Transition forest/ savanna	Stover, vines, groundnut cake

Farming system	Major crops	Major animals	Main regions[a]	Feed source
Highland agriculture (*continued*)	Millet/sorghum maize, groundnuts, cowpeas, sesame, tobacco, cotton, vegetables, cassava, yams	Cattle, goats, poultry, sheep, horses, donkeys	Savanna (Guinea and Sahel)	Crop residues, some oil cake, brans, stover, vines, cull tubers
Flood land and valley bottom agriculture (animals of some importance)	Rice/yams/plantains, maize, vegetables, sugarcane, rice, cocoyams, millet, groundnuts	Goats, poultry	Humid tropics	Crop residues, vines, grazing
	Rice, vegetables, maize, millet, groundnuts, plantains, sugarcane, cocoyams	Cattle, goats, poultry, sheep, swine, horses, donkeys	Transition forest/savanna	Straw, stover, molasses, brans, groundnut cake
	Yams/sugarcane, maize, cowpeas, cocoyams, groundnuts, vegetables, plantains, rice, yams	Cattle, goats, sheep, poultry, swine, horses, donkeys	Savanna (Guinea and Sahel)	Vines, brans, tubers, molasses, sugarcane tops
Mixed farming (farm size variable, animals important)	Rice/yams/plantains	Two or more species (widely variable)	Humid tropics	Fallow, straw, brans, vines
	Rice/vegetables, yams, cocoyams	Some cattle	Transition forest/savanna	Fallow, vines, straw
	Sorghum/millet, groundnuts, cotton, tobacco, maize, cowpeas, vegetables	Cattle, goats, sheep, poultry, horses, donkeys, camels	Savanna (Guinea and Sahel)	Stover, vines, fallow
Plantation crops, East Africa (small holdings, animals of some importance)	Coconuts, vegetables, maize, plantains, cocoyams, cassava	Cattle, horses, donkeys	Humid tropics Transition forest/savanna	Herbage from grazing or cut and carry
Plantation crops (compound farms,[a] etc.) (animals of some importance)	Cacao, vegetables, maize, plantains	Goats, sheep, poultry, swine	Humid tropics	Herbage from grazing or cut and carry, stover
	Tree crops, sugarcane, plantains	Goats, sheep, poultry, swine	Transition forest/savanna	Herbage from grazing or cut and carry, sugarcane tops
Market gardening (animals may or may not be present)	Vegetables[a]	Variable	Humid tropics Transition forest/savanna	Natural rangelands, crop residues, browse plants, range forbs

continued

Table 8.2 (continued)

Farming system	Major crops	Major animals	Main regions[a]	Feed source
Latin America Perennial mixtures (large farms) (livestock relatively unimportant)	Coconuts, coffee, cacao, plantains, bananas, oil palm, sugarcane, rubber	Cattle, swine	All[c]	Natural pastures, by-products, cull material
Commercial annual crops (medium to large farms, livestock moderately important)	Rice, maize, sorghum, soybeans (on small farms)	Swine, cattle, poultry	All except CI[c]	Pasture, crop residues, grain
Commercial livestock *Extensive* large to very large (livestock dominant)	None are important	Cattle (beef)	C, V, Br, Bo, G, CA[c]	Natural grasslands
Intensive Medium to large, livestock dominant	Improved pasture, some grains	Cattle (dairy), swine, poultry	All[c]	Natural and improved pasture, feed grains, by-products
Mixed cropping, small size in settled areas; medium size in frontier areas; subsistence or cash economy (livestock relatively important)	Rice, maize, sorghum, beans, wheat, cacao, plantains, coffee, tobacco	Cattle, poultry, goats, sheep, donkeys, horses, mules, swine	All[c]	Natural pastures, crop residues, cut feed

[a] Enclosed areas around household or villege.
[b] Present or absent, depending on the area.
[c] All, all countries; Bo, Bolivia; Br, Brazil; C, Colombia; CA, Central America; CI, Caribbean Islands; E, Ecuador; G, Guyana; P, Peru; V, Venezuela.
Underlining indicates the main system.
Source: McDowell & Hildebrand (1980).

8.3
Grassland research

Traditional attempts to solve grassland problems have taken a reductionist approach (Fig. 1.1). However, there are problems relating to the management of grassland systems which are not amenable to reductionism. This is particularly true for grassland problems which involve (i) social and organizational components within the system and (ii) 'externalities' which are not easily quantified: factors outside the farming system such as the values which urban-dominated societies place on conservation and the sustainability of agriculture. Systems research which places emphasis on these 'soft' components of farming systems is evolving to complement reductionist research (Shaner, Philipp & Schmehl, 1982; Norman & Collinson, 1985; Bawden *et al.*, 1985; Miller, 1985).

The problem of acidification of grassland soils illustrates how resolution of some grassland problems depends on their being seen within the context of a farming system. The biological causes of soil acidification in legume-based grasslands are complex (Chapter 5). Reductionist research is providing an understanding of the chemistry and biology of the problem; technological research has shown that application of lime will temporarily reduce soil acidity and ameliorate the problem. This in itself does not 'solve' the problem. A 'hard' systems approach can place the need for lime in a whole-farm context and determine biological and economic optima for the application of lime for a particular farming system (e.g. Hochman *et al.*, 1987). Alternatively the hard systems approach may be used to engineer new systems using data generated from reductionist research or to highlight deficiencies in existing systems, thus providing research direction. Reductionist models of lime application give radically different predictions depending on the ratio of current costs and prices. For example, in Australia it is currently economic to apply lime to grassland only via a cropping phase. This outcome necessitates changes in farming practice and provides impetus for research on new crop–pasture systems. Where cropping is not possible or where markets change or do not exist for feasible crops, the question becomes whether solution of the acid soil problem is left open to market forces (the cost of lime and other inputs versus the returns from farm outputs and thus farmers' terms of trade) or whether because of the potentially damaging long-term consequences of not applying lime, another approach is required.

The possible long-term effect of acidification is the non-sustainability of pastoralism, as illustrated simply in Fig. 1.4d. This reintroduces the concept of sustainability and the value society places on the sustainability of one of its resources. At this level the formulation of a policy to improve acid soils is complex and ambiguous and there is seemingly no clear right answer because we do not yet know how to quantify social values in a way which is strictly compatible with decision-making for biological systems. Such situations are amenable to analysis using a soft systems approach (Fig. 8.2). Using this approach with the acid soils problem, grassland systems analysts, together with farmers, researchers, extension officers and resource planners would attempt to communicate different perceptions of the situation as a whole. First they would be required to establish the boundary of

Fig. 8.2. Grassland research methodologies. When a problem is experienced various cyclical methodologies exist to resolve it; they are not mutually exclusive and within one problem the researcher may, or may not, move between different levels in this 'spiral'. (Adapted from Bawden, 1985.)

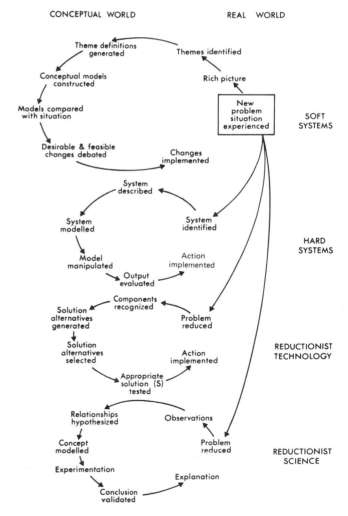

the system under consideration, i.e. whether the system is a farm, a region, a state or a nation. The use of systems thinking would be required to generate a conceptual model of the system, which is then compared with the real (as opposed to conceptual) situation which has previously been analysed (Wilson, 1984). The insights this process provides are the basis of a debate about desirable and feasible change among people involved with the problem.

Grassland agronomists frequently confront complex issues such as defining research priorities or developing more efficient systems of germplasm introduction, evaluation, cultivar release and adoption. Both reductionist research and the use of 'soft' systems methodology (Checkland, 1985) or agro-ecosystems analysis (Conway, 1985a, b) have continuing roles for innovation. Fig. 8.2 presents such a view of reductionist and systems research methodology. Both methodologies are cyclical, i.e. the researcher returns continually to observe and try to improve the grassland system. The reductionist and systems approaches are not mutually exclusive: the choice of approach depends on the nature of the problem. It may also be desirable to iterate between approaches in the research process.

8.4
Environment

Grasslands in farming systems are subject to weather and climatic variation. This variation constrains productivity and leads to farm management policies aimed at coping with experienced or anticipated variation. Even in the seemingly well-endowed climate of southern England the variation in the annual yield of grassland ranges from 40 to 160 per cent of the mean annual yield (Corrall, 1978). Matching the supply of feed with animal demand is difficult in the face of this variation. Farmers choose stocking rates which are often justifiably conservative and these account for most of the potential productivity 'lost' from grasslands.

Natural variation, especially in rainfall, also hampers many management operations related to grazing and conservation. This reduces the utilization of existing grassland. Of greater concern in some systems (e.g. semi-arid areas and rangelands) are major perturbations such as are caused by drought, fire and flood.

The science of grassland agronomy evolved slowly to cope better with the effects of natural perturbations. The evaluation of new germplasm was for many years characterized by measuring only productivity: the amount of plant dry matter produced per season or year. The use of grazing experiments and the introduction of animals at early stages of pasture plant evaluation to assess acceptability, grazing resistance and animal production has helped agronomists to focus on the quality as well as the quantity of grass being produced. Perhaps more importantly the poor performance of many introductions in commercial practice, and the management difficulties associated with maintaining grass–legume swards, has led grassland agronomists to focus on persistence and survival (e.g. Humphreys, 1984; Jones, 1986). This has required eco-physiological studies to understand what Humphreys (1981) describes as 'pathways to persistence'. An increasing focus on forage trees is also, in part, a response to their better persistence and productivity in the face of climatic variability (Brewbaker, 1986).

The focus on persistence and stability of yield has been greatest in the more climatically variable tropical and subtropical regions and in mediterranean regions which have relatively low year-to-year variability but extreme environmental stress within each year. In these areas continuing emphasis on increasing productivity may be at the expense of pasture quality; this focus will probably change owing to changing cultural forces (Section 8.8) and increasing emphasis on legumes.

8.5
Evolution

Grasslands have many of the properties of natural ecosystems; many of the world's natural grasslands are managed to maintain them in a sub-climax state. These grasslands usually comprise a wide range of species. In Chapter 1 the concepts of succession, stability and sustainability were introduced. The effect of man's intervention in grassland management can often be at odds with natural or homeostatic mechanisms and so be vulnerable to evolutionary forces.

Increasingly, grassland species are being sown outside the areas in which they evolved, i.e. their centres of diversity (Harlan, 1983). They have been collected by man in their natural habitat, often as small quantities of seed from single plants, and introduced into other regions or countries where they do not occur naturally. The species is then evaluated under climatic, biotic, edaphic and socio-economic conditions characterizing research organizations and farming systems in the foreign region. This germplasm is now finding its way into international seed distribution networks (e.g. Toledo, 1986a). While this global distribution of germplasm has the potential to improve grassland and livestock production in many farming systems, there are potential dangers. These dangers are best considered from an evolutionary viewpoint (Lowrance, Stinner & House, 1984).

One important example of evolutionary forces and dangers is the narrowness of the genetic base of some modern pastures. Many improved pastures exist as monocultures of a single species or, in some cases, a single ecotype. In other cases they may comprise mixtures of one grass and one legume. Pastures with such a narrow genetic base are highly vulnerable to pathogens and predators (Lenné, Turner & Cameron, 1980). The examples of lucerne cv. Hunter River attack by aphids and *Stylosanthes humilis* decimation by anthracnose in southern and northern Australia respectively were mentioned in Chapter 1. Fall armyworm (*Spodoptera frugiperda*) is a major pest on exotic, i.e. imported pastures of coastal bermudagrass (*Cynodon* sp.) in south-eastern United States as are the *cigarrinhas* (spittlebugs) on *Brachiaria decumbens* grasslands in South America. Martin (1983) argues that while the *cigarrinhas* problem in Brazil is of tremendous magnitude, it is only a symptom of the problem of large monocultures of *cigarrinha*-susceptible *B. decumbens*, which produce an overwhelming inoculum of *cigarrinha* for other, less susceptible grasses such as *Panicum maximum*. The browse leguminous shrub *Leucaena leucocephala*, which is now almost pan-tropical and with a limited genetic base, is also in danger of decimation by a sucking bug related to aphids (*Heteropsylla* sp.); the bug originates from *Leucaena*'s centre of diversity in Central and South America.

Natural grasslands rarely suffer from insect pest pressure though cyclic insect problems may occur. Grazing, fire and species diversity are the means of moderating the effects of pest and pathogen attack. Much of man's ability to use these as tools of management has been lost due to changing land use and social values. It has also been aggravated by the introduction of exotic, fire-sensitive grassland species. Future strategies should include making more specific plant selections to achieve diversity (Martin, 1983) and the desired regulation of pests (Burdon, 1978). Thus the maintenance of genetic diversity in itself becomes a criterion for new pasture plant release. In this regard maintenance, rather than complete eradication, of a diverse weed flora may also be advantageous (Martin, 1983). Lenné *et al.* (1980) also argue for initial germplasm evaluation in the original centre of diversity as this is also the centre of diversity of associated pests and pathogens. This would reduce the risk, inherent in the global movement of seed, of the accidental introduction of a virulent pest or pathogen into a pest- or pathogen-free area.

8.6
Economics

Reviews of socio-economic constraints on grassland production systems in the United Kingdom (Jollans, 1981), Europe (van Dijk & Hoogervorst, 1983), South America (Toledo, 1986*a*, *b*) and South-East Asia and the Pacific (Perkins *et al.*, 1986) reveal the potency of these forces and the frequent inability of traditional production-orientated research to cope with them. In the United Kingdom, for instance, where grasslands occupy 70 per cent of the agricultural area, socio-economic projections made in 1980 refuted the enthusiasm of grassland agronomists for ever more output and increasing intensification of production regardless of cost (Jollans, 1981, p. 539). It was argued that:

   (i) increased output could only be justified if cheaper livestock product prices ensued;

   (ii) aggressive marketing policies for the output from grasslands were required; and

   (iii) substantial areas of grassland (up to 9 per cent of current area) might be released for amenity and other purposes.

This last argument reflects changing cultural values (Section 8.8). Together these arguments appear forceful given that technology is already available that could more than double the output of UK grassland.

In tropical South America the nature of the socio-economic environment has determined the research strategy being pursued. Here, research aims to produce '. . . a low-cost, low risk grassland technology that will persist and produce in times of national, regional or individual negative socio-economic conditions, and will spread and increase productivity under favorable conditions' (Toledo, 1986*a*). Significant gains in productivity have been achieved with this technology. For example, in the Llanos of Colombia, planting small areas (5.6 per cent of the ranch) of improved species caused the liveweight gain per animal to double and the gain per ha to increase ten-fold when animals were grazed on the improved pasture compared with native savannah (CIAT, 1983). This resulted in impressive (30 per cent higher) marginal internal rates of return on capital invested (Table 8.3).

Productivity increases and economic gain from developed grassland technology are unfortunately not always sufficient for ensuring success. Lack of success raises two important issues: the forms of economic analysis available and the so-called 'adoption' of new technology.

In economic analysis, pasture is rarely traded as a commodity, although pasture sold as hay, silage or green chop, including hand-cut and carried feed are exceptions. The value of pasture as a feed must therefore be imputed from its contribution to livestock production, or as is sometimes the case, especially with nitrogen-fixing legumes, to the succeeding crop in a rotation. Thus the economic value of pasture and even value-added pasture products (e.g. hay) is subject to price changes characteristic of livestock commodity markets and consumer demand. These perturbations may impact at local, regional, national or international levels depending on the position of the livestock sector in a region or a nation's economy (e.g. Marsh, 1981) and the extent of dependence on and access to local and international markets. This makes economic analyses more difficult than in, say, cropping systems.

Forms of economic analysis depend on purpose; some purposes include:
(i) to make an investment decision on an 'improvement' to existing technology or to differentiate between a range of available techniques within the 'technology';
(ii) for grassland agronomists, to gauge the likely economic implications of their research findings;
(iii) at the national or macro-level, the need to consider the contribution of grassland production to sectoral (e.g. beef, wool) productivity; and increasingly
(iv) to assess the contributions of grassland to land stability and erosion control (e.g. Cox, 1984).

The last-mentioned contribution is less able to be quantified and economic methodologies, which may or may not be appropriate, are only now being developed.

Table 8.3. *Biological and economic impact five years after planting 5.6 per cent of a Llanos ranch in Colombia with* Andropogon gayanus *(40 ha),* A. gayanus–Stylosanthes capitata *(80 ha) and* Brachiaria decumbens–Desmodium ovalifolium *(25 ha)*

	Year		Increase (%)
	1979	1983	
Biological			
Cow weight (kg lw)[a]	255	328	29
Weaning rate (%)	50	57	14
Weaning weight (kg lw)	109	162	49
Stocking rate (AU/ha)	0.13	0.24	85
Economic			
Marginal internal rate of return			31

[a] lw, liveweight; AU, animal units.
Source: Toledo (1986*b*).

For a farmer concerned with investment in a technology (new species, fertilizer, cultivation and/or sowing equipment, etc.) which may be a package, or more usually part of a 'technology package' adapted to individual farmer circumstance, there is a large initial outlay of capital with returns spread over a number of years. The farmer must have accumulated cash reserves (savings) or have access to credit to finance the initial investment and to service interest payments if these cannot be met from normal or increased cash flow. In this example 'pasture improvement' is not unlike any on-farm development project and the best available form of economic analysis is discounted cash flow analysis which takes account of differences in the timing of income and expenditure. A worked example is presented as Appendix C.

Investment in improved grassland technology is primarily made to increase productivity. However, increasingly there is likely to be interest in stabilizing productivity at an acceptable level by reducing season-to-season or year-to-year variability. Potential economic effects from increased grassland production interests the farmer, the grassland researcher and resource economists; the likely economic effects include:
(i) an increase in the stocking rate;
(ii) a reduction in the dependence on alternative feeds;
(iii) a reduction in the need to buy in fodder (e.g. hay) or rent additional grassland;
(iv) the release of land for alternative enterprises;
(v) a decrease in the time to produce a given animal product;
(vi) an increase in reproductive performance (e.g. conception, calving or weaning rate; return-to-service); and
(vii) a reduction in the time spent cutting and carrying forage.

Doyle & Elliot (1983) present methodologies for the economic evaluation of (i)–(iv). As an increase in the stocking rate is possibly the most common outcome of an increase in grassland production, their methodology for evaluating this is outlined below:

$$\Delta GM = gm \times \Delta N \qquad (8.1)$$

where ΔGM is the change in the gross margin ($ per ha), gm is the gross margin per animal ($ per head) and ΔN is the change in the stocking rate (head per ha). The magnitude of ΔN will depend on both the absolute increase in grassland yield and its seasonal distribution. In the simple case where the absolute increase in the stocking rate is proportional to the absolute increase in the production of dry matter, ΔN may be defined as:

$$\Delta N = (\Delta Y_G / R_G) \times U/100 \qquad (8.2)$$

where Y_G is the increase in annual grassland production (kg DM per ha), R_G is the annual feed requirement per animal (kg DM) and U is the efficiency of utilization of the extra feed (per cent). Substituting Eqn (8.2) in Eqn (8.1), the average increase in gross margin in cents per kg DM of extra feed produced (V) is given by:

$$V = 100 \times (\Delta GM/Y_G) = (gm/R_G) \times U \qquad (8.3)$$

This equation takes no account of increased investment costs (I) incurred through the purchase of extra stock, the provision of housing or the payment of associated interest. Thus the true financial benefits from an extra kg DM of feed (V) are:

$$V = \frac{(gm - I)}{R_g} \times U \qquad (8.4)$$

The value of V will vary according to specific assumptions about prices, costs and grass utilization rates and takes no account of changes in labour and machinery requirements. Doyle & Elliot (1983) further emphasize that if there is a marked seasonality in grassland yield improvement due to the technology, then ascribing a single value to ΔY_G is inappropriate. For example, extra production has no economic value when confined to a period when feed supply already exceeds demand.

When investing in new technology the particular strategy to choose may not be clear to a farmer, e.g. the choice of the method of pasture establishment. For making such decisions Gruen (1959) identifies only two narrowly-defined conditions when emphasis should be placed on maximizing the amount of production per ha:

(i) when improvable land is the factor limiting production growth; and

(ii) when the most intensive technique maximizes not only the production per ha but also the production per unit of other scarce resources (e.g. money and labour), i.e. the cheapest method per unit increase of (discounted) net income.

The most limiting resource will vary from farm to farm and as a consequence a farmer may not be interested in the cheapest method (per unit increase in output or income) but in the method which produces the largest income per unit of some scarce resource, e.g. available money. Modelling, e.g. linear programming, can help in this evaluation and decision-making process (Section 8.11.2).

The economics of labour is another consideration in terms of both cost and availability. In some grassland systems the contemporary emphasis is on labour-saving methods; e.g. one pastoralist in southern

Australia can care for 5000 sheep (Morley, 1986). In systems where labour is readily available or underemployed, the introduction of grassland technology may still be constrained by labour due to impositions made on the farmers' lifestyles. Changing socioeconomic circumstances, such as the increasing educational opportunities for children in some third world societies which result in the loss of traditional livestock managers, also impact on the economic feasibility of improving grassland production.

8.7
Technology

The development of biologically-feasible technologies for grassland improvement does not ensure their use by farmers or their 'transferability' from one farming system to another. The major barriers are variation in farmer/manager purpose, economic perturbations and cultural forces. In the South American example quoted earlier the costs for the basic inputs of developed grassland technology vary widely (Fig. 8.3). This constrains the adoption and transferability of all but exceptionally viable or 'robust' technologies.

In South-East Asian and Pacific farming systems Perkins et al. (1986) identify many opportunities for the production of forages. The realization of these opportunities is currently constrained, not by physical and biological factors, but by the farmer's goals and attitudes to stock rearing and access to forages. Four major barriers to forage improvement are:

(i) access to land and its product;

(ii) the complexity of forage management;

(iii) markets and the marketing of the animal product; and

(iv) farmer motivation.

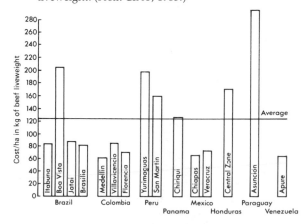

Fig. 8.3. Relative cost of a package of basic inputs for the establishment of 1 ha of improved grassland throughout Central and South America, in terms of kg beef liveweight. (From CIAT, 1983.)

In a human activity system comprising usually a rural household, changes and suggestions will be adopted if they meet some perceived need, they assist the farmer or farm household to achieve some goal, they are sympathetic to the individual situation and they are viewed as achievable.

Options for technological change are thus evaluated by the farmer within the constraints of economics and culture. Furthermore, an adoption leads to further opportunities for technological adoption; the open-ended choices resulting from technological change may be visualized as a 'decision tree'. That is, any decision to adopt new technology can be visualized as a yes or no choice and once made, that choice leads to a series of yes–no choices. For example, if a decision is made to remove trees, that raises the choice as to whether grasslands should be sown, fertilizer applied, pests controlled, fences erected, etc.

Given that technological change is evaluated by the farmer on cultural criteria as well as on the basis of economics and the feasibility of options which arise once the first choice has been made, it is important to understand the management, as well as the biological, aspects of a farming system before developing technology for that system. Understanding arises from surveys and rapid appraisal (Beebe, 1985). For example, Basuno & Petheram (1985) described many important but non-biological aspects of farming systems in West Java, Indonesia:

(i) Of the total village area 10 per cent comprised road reserves, 40 per cent house yards and 50 per cent cultivated land.

(ii) Forage rights on roadsides, banks and canals were communal and crop residues were usually available on request from the farmer.

(iii) Field and perennial crops were considered more profitable than livestock.

(iv) No crops were grown specifically for feed and forage. Animal rearers relied heavily on crop residues of cassava, sweet potato and legume tops, rice straw and rice bran. The villagers thus had no 'cultural experience' of sowing pasture to feed animals.

(v) There were a range of reasons for keeping livestock.

(iv) Animals were mainly hand fed; ruminants were fed entirely on forage and were only grazed when forage was scarce.

(vii) Households rearing ruminants (sheep and buffalo) were greatest in number in rural or traditional hamlets, yet goat density was highest in more urban hamlets along the main roads.

(viii) There were significant variations in land and animal ownership patterns.

(ix) Fewer rearers of ruminants (20 per cent) had attended school than had chicken rearers or non-livestock rearers (57 per cent).

8.8
Culture

Cultural forces take different forms in different societies but they may be pervasive in shaping the future directions of livestock industries and thus the grassland systems on which these depend, in constraining the adoption of technology developed without an appreciation of these forces, and in reshaping the use to which grassland plants, and thus grassland agronomists working with these, may be put. McKenzie (1981) and Hollingsworth (1981) review respectively the impact of changing consumer demand and the cultural awareness of nutrition on British grasslands and their products. There is a general concern about the fate of the agricultural sector in many economies and a growing concern for developing sustainable agricultural systems (Douglas, 1985); grassland systems are likely to play a greater role in this context, particularly in rotation with crops and for soil conservation strategies.

Asian farming systems have complex cultural patterns associated with livestock industries. Christianity and Islam have no taboos on meat-eating but Buddhism and Hinduism do, although increasing urbanization is leading to cultural changes in the latter group. Different types of animals have different rankings or status in different regions; animal sacrifice is common, and the status of livestock keepers varies in different countries (Wimaladharma, 1985).

Cultural patterns have often developed for good reasons. For example, the productivity per h of semi-nomadic livestock farming systems in the Sahel is greater than that of sedentary systems in the same area and of comparable regions in Australia and the United States (Table 8.4). The semi-nomadic system of farming is complex and suited to the biological and socio-economic conditions. It is able to capitalize on regional variations in forage quality with inbuilt checks (such as lack of water) to limit over-exploitation (Fig. 8.4). Development options for improving the situation of the Sahelian animal farmer are not straightforward: Breman & Wit (1983) provide a good analysis of the dangers that might ensue from uninformed intervention in farming systems.

8.9
Integration of grasslands and crops

Grasslands are inextricably linked with crops and their residues in many farming systems. Grassland may be intercropped or sown as pure stands in a cropping sequence for one or more years. Grassland,

Fig. 8.4. Seasonality of weather, herbage availability (pasture dry weight) and protein concentration (per cent) and animal migrations in the semi-nomadic farming system of west Africa, the Diafarabe herd in Mali. (From Breman & Wit, 1983.)

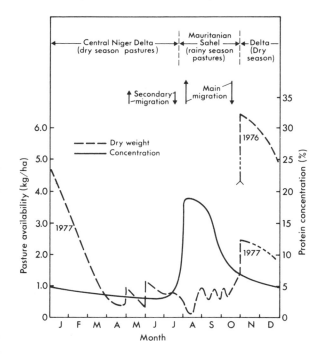

particularly improved pasture, when integrated into cropping systems has the potential to:

(i) Replenish soil fertility, particularly with respect to nitrogen, from legume dinitrogen fixation and the more rapid cycling of nutrients via returns of dung and urine of the grazing animal. The contribution of varying years of growth of legume-based pasture in a wheat–pasture system of southern Australia to total soil nitrogen is shown in Fig. 8.5a. Legume pastures have the potential to contribute significantly to subsequent grain yield, as shown in Fig. 8.5b.

(ii) Improve soil structure and stability. This is demonstrated by an increase in the percentage of water-stable soil aggregates during a pasture phase of varying years in contrast to a decline in water-stable aggregates and thus in soil structure associated with continuous cropping of wheat (Fig. 8.5c). In addition dead pasture mulch can alter the soil surface radiation balance so as to retard rises in soil temperature and soil strength by slowing drying (Fig. 8.5d). These changes enhance crop germination.

(iii) Break a pest or disease cycle. This is not always effective, as reviewed by Reeves (1986).

(iv) Provide better weed management for the current or subsequent crop. This is not simple as in many systems valuable pasture species may be weeds at other times. Intercropping of cassava (*Manihot esculentum*) and maize with annual or perennial tropical legumes of the genus *Stylosanthes* has been shown to control weeds and provide additional forage for livestock without any detrimental effect on crop yield in both Asian and African farming systems (Humphreys, 1981).

Table 8.4. *Livestock production in the Sahel and two comparable regions (semi-arid tropics with less than 500 mm of rainfall per year)*

Region	Protein production (kg/year)		Fossil energy input (10^3 MJ/man-hour)
	Per ha	Per man-hour	
United States	0.3–0.5	0.9–1.4	105–146
Australia	0.4	1.9	628
Sahel			
Nomadism	0.4	0.01	0
Transhumance[a]	0.6–3.2	0.01–0.07	0
Sedentary	0.3	0.04	0

[a] Seasonal movement of livestock from one defined region to another.
Source: Breman & Wit (1983).

Competition from the otherwise useful pasture grass *Lolium rigidum* depresses wheat yields in southern Australia.

(v) Supplement and upgrade the quality of crop residues or native grasslands; intercropping of crops such as maize, sorghum, cassava and kenaf with pasture legumes has been shown to be feasible. The legume may be harvested during crop growth or after crop harvest to supplement crop by-products or the lower quality crop residues. Legumes may also be oversown into maturing crops to utilize residual soil moisture (Shelton, 1980), on crop borders or rice paddy walls as in north-east Thailand (Gutteridge, 1983) and as alleys of browse shrubs.

(vi) Increase the efficiency of land use under plantation crops. Young crops of coconut, rubber, oil palm and some fruit and nut trees provide a suitable light environment for pasture growth. Animal production from grazed legume–grass pastures under coconuts is not detrimental to coconut yield; pastures reduce competition from weed species, obviate the need to slash and control growth and help in locating fallen coconuts. Plucknett (1979) and Rika (1986) review this subject in detail. Pasture life and productivity is reduced as the level of shading increases when stands age or where tree density is too high. Agroforestry is a special form where trees are planted at lower densities to permit long-term integration with grazing (Heady, 1986).

(vii) Maintain the nutrition of draught animals; in some intensive irrigated cropping systems of Asia, e.g. rice, land for grazing and forage for draught animals is often more limiting in the wet or cropping season. This may be alleviated by sowing paddy walls, field borders and communal grazing lands with improved forages as discussed above.

Fig. 8.5. Effects of grassland in grassland–crop systems on soil fertility, structure and temperature. (a) Changes in levels of total soil nitrogen following years of pasture (P) initially sown under wheat (U) and wheat crop sown alone (W) in northern Victoria, Australia. U/sown is wheat sown at the same time as pasture (i.e. undersown) in the same paddock with the seed mixed together. (b) Crop (maize) yield response to applied nitrogen fertilizer following one year of legume or grass pasture at Katherine, northern Australia. (c) Changes in soil structure, as indicated by water stability relative to soil under continuous wheat; P, U, W as for (a). The percentage of water-stable aggregates in soil sown to continuous wheat is taken as the base (100%). (d) Effects of various amounts of legume mulch (0–1900 kg per ha) on surface soil temperature measured at 1500 h at 5 cm depth at Katherine, northern Australia. (From: (a) and (c) White *et al.*, 1978; (b) and (d) McCown, Jones & Peake, 1985.)

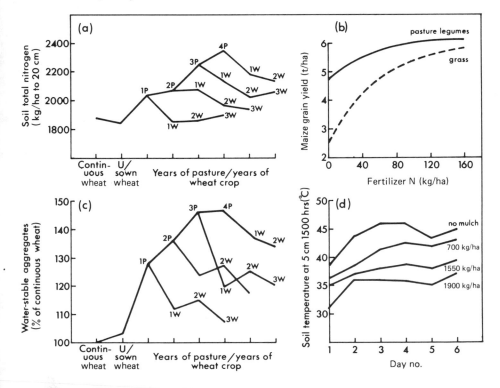

(viii) Provide enterprise and thus economic diversification by incorporating animal production into cropping systems (see IRRI, 1983).

The extent of the interactions in animal–pasture–crop systems is demonstrated in Fig. 8.6. This is based on a proposed farming system for northern Australia which integrates the common practice of extensive beef production on native grasslands with a legume pasture, pasture intercrop and maize and sorghum cropping. The region has a history of failed cropping systems (Muchow, 1985) and the proposed system seeks to develop a sustainable crop–animal farming system for the region.

Introduction of cropping into permanent grassland has the potential to enable the grassland to benefit from the crop by utilizing residual fertilizer (e.g. phosphorus) applied in the more economically viable cropping phase. The crop phase also permits economic incorporation of lime or gypsum for amelioration of acid soils or soil physical problems. Also, the crop pre-empts invasion of the pasture by nitrophilous weeds as it utilizes nitrogen built up under leguminous pasture.

Integration of crops and grasslands adds to the complexity of the farming system. Clewett, McCown & Leslie (1986) point out that gains from integration

need to be substantial to warrant the use of a more complex system.

8.10
Alternative uses of grassland plants

The expertise of agronomists is increasingly likely to be directed towards the development, management and utilization of grassland plants in non-traditional situations.

The development of alternative uses for grassland plants will depend on:
  (i) changing demands for, and supply of the products of grassland;
 (ii) changing farmers' terms of trade with respect to input costs;
(iii) technological developments in grassland production and utilization; and
(iv) trends in socio-economic and cultural forces such as increasing leisure, concern about soil erosion and concern about the sustainability of the environment.

The potential range of alternative uses for grassland plants is likely to be determined by two not mutually exclusive factors. These are the chemical and physical properties of the plants and the properties of the plants in communities.

Fig. 8.6. A grassland–crop farming system being evaluated at Katherine, Australia, for the wet-and-dry tropics. (Adapted from McCown *et al.*, 1985.)

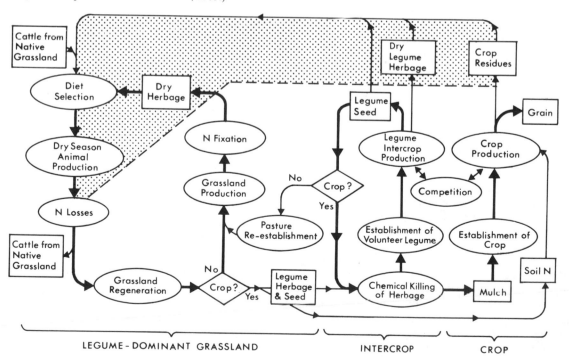

8.10.1
Uses based on the chemical and physical properties of plants

The chemical composition of grassland species was discussed in Section 6.1. The chemical properties of plants determine the potential utility of individual plants in the schema presented in Fig. 8.7, which describes some present and possible future uses of grassland plants. The physical properties of plants include the energy content and the ease of fractionation of the plants (extracting sap or juices). In the schema given in Fig. 8.7 fresh pasture is subjected to wet fractionation by:

  (i) cell rupture (maceration or pulping);
 (ii) expression of the juice;
(iii) the separation of protein concentrate from the juice fraction; and
 (iv) the storage and preservation of products, including pressed herbage, which usually has a moisture content of 65–75 per cent and which by further hydrolysis may yield simple sugars.

Jorgensen & Lu (1985) review the wet fractionation of legume forages and recognize the need for further research on the development of processing equipment that is energy efficient and low cost, the removal of antiquality factors (Chapter 6), the preservation of protein concentrate and the processing of deproteinized juice.

The high levels of crude protein, and thus nitrogen, in pasture or forage legumes has long been recognized in their use as green manure crops. More recently browse species have been utilized in alley-cropping systems where legume leaf is a source of additional nitrogen and organic matter for the inter-crop.

Grassland plants may also be used as fuel, as recognized in the schema in Fig. 8.7. Many browse species such as *Leucaena* may fulfil a dual purpose in providing both fuel and forage. It may also be possible to use green pasture in the production of biogas (Küntzel, 1982).

The physical properties of some species, e.g. *Imperata cylindrica*, make them suitable for use in thatching and other types of roofing. Others, e.g. *Cynodon dactylon*, are reputed to have medicinal properties.

8.10.2
Uses based on the properties of grassland communities

Grassland communities may be used for a diverse range of present and potential purposes. Grassland plants may be used to stabilize land after soil disturbance, as in mining reclamation; for soil conservation purposes; for weed control purposes as in some plantation cover crops; for leisure and aesthetic purposes (e.g. Tranter & Tranter, 1981) and for grazing by native animals, as in national parks.

Conversion of forest to grassland or crop generally increases the quantity of surface water run-off. Conversion of native perennial grasslands to improved annual pastures of *Lolium rigidum* and *Trifolium subterraneum* has reduced run-off in southern Australia but the perennial grass *Phalaris aquatica* is recognized as being the most stable for these purposes (Mitchell & King, 1980). The use of grassland for erosion control depends on maintaining ground cover. Cover reduces the effects of raindrop impact on the soil by reducing splash erosion: rates of soil detachment and aggregate break-down are reduced, aggregate structure is maintained and there is less surface crusting or sealing so that infiltration rates remain high and surface run-off is reduced. Soil erosion decreases either linearly or exponentially with increasing cover (Morgan, 1985). Problems such as the development of soil acidity under legume-based grasslands may reduce ground cover and limit the effectiveness of the grassland in erosion control.

8.11
Modelling of grassland systems

The word 'model' and the act of 'modelling' have many meanings. Here we use a very broad interpretation of Wilson's (1984): 'a model is the implicit interpretation of one's understanding of a situation, or merely of one's ideas about the situation. It can be expressed in mathematics, symbols or words, but is essentially a description of entities and the relationships between them. It may be prescriptive or illustrative, but above all it must be useful.' Wilson (1984), following Ackoff (1962) recognizes four forms of models:

  (i) conceptual
 (ii) analytic
(iii) analogic; and
 (iv) iconic.

Conceptual models include pictorial or symbolic models which cover qualitative aspects of a situation. The other forms are physical (iconic and analogic) or quantitatively formulated (analytic) models. We are concerned here with conceptual and quantitative modelling.

Modelling is carried out for some purpose. Rykiel (1984) recognizes four purposes (Table 8.5). This reflects the notion that a model is not a goal in itself but rather a part of some analysis associated with problem resolution which leads to increasing knowledge and an increasing need for quantitative results. Different forms of modelling find greater use in different problem-solving strategies and at different levels of biological organization (Fig. 1.1). Used in this way

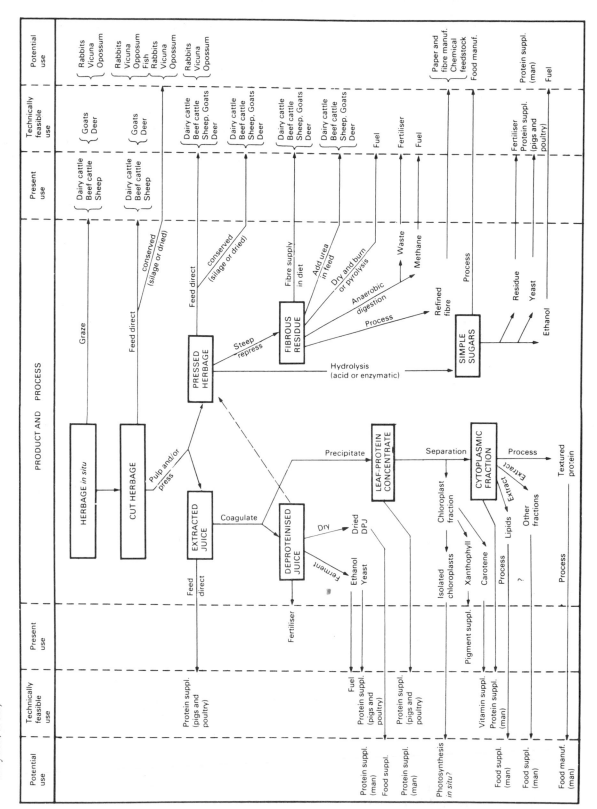

Fig. 8.7. Schema describing present and possible future uses of herbage, both whole and fractionated. (From Snaydon, 1981.)

modelling may serve as a unifying function in the study of grasslands in farming systems if the steps outlined in Table 8.5 are followed within a relevant systemic or multidisciplinary framework (Rykiel, 1984).

8.11.1
Conceptual modelling
Conceptualization of what the system under study is or might be is usually a starting point (Rykiel, 1984; Wilson, 1984). Thus conceptual modelling precedes other forms, as well as being a modelling form in its own right.

Conceptual or qualitative models may be used:
 (i) as an aid to clarifying thinking about an area of concern;
 (ii) as an illustration of a concept;
 (iii) as an aid to defining structure and logic; and
 (iv) as a prerequisite to design.

These uses are not mutually exclusive (Wilson, 1984). We have already used examples; Fig. 2.2 combines aspects of (ii) and (iii), presenting some of the key concepts and interactions in plant generation. Spedding (1975a) has devised a circle diagram form of a conceptual model. This defines sub-systems which can then be used to investigate the effect of, say, the amount of grass grown in a sheep production system on profit (Fig. 8.8).

The modelling of human activity systems, as distinct from biological or natural systems, utilizes a particular form of conceptual model. The modelling language developed by Checkland (1981) and colleagues is the use of *verbs* to model functions or activities. Plant domestication can be viewed as a human activity system (Fig. 8.9). Within each sub-system denoted by a verb to describe the main function or purpose, occur various organizational and research activities. Each sub-system may be broken down to further sub-systems to gain greater insight into the range of functions involved, i.e. the concept of hierarchy. By comparison of the function or the 'what', e.g. commercialization, with a range of 'hows', e.g. farmer cooperatives, appropriate structures or processes can be identified for each system or sub-system.

Table 8.5. *Modelling: the purposes for which a model might be used and a four-step modelling procedure which enables the development of a common perspective by different discipline groups involved in problem-solving or research in connection with pastoral systems*

Purposes

1. Exploration	Objectives are very often general or intuitive, usually with no specific criteria for meeting them; the main aims are insight, clarification and understanding of the factors that contribute significantly to system behaviour
2. Explanation	The general objective is to understand the structural and functional relationships between components and sub-systems that explain the pattern of inter-connections within the system and generate system behaviour; specific objectives are related to the level of resolution and the level of the study: system, sub-system, and component levels
3. Projection	The objective is to examine the dynamic behaviour of system variables at any level (i.e. component, sub-system or system) and the effects of changes in the values of parameters or variables and their variability; variability in the occurrence of events and making of decisions; the patterns of behaviour represented in the dynamic relationships among variables are more important than the actual values of the variables
4. Prediction	The specific objective is to estimate future values of particular system variables and/or the nature and timing of events and decisions; emphasis is on the accuracy and utility of the prediction, and the reasonableness of the explanation of the prediction

Procedures

1. Formulation	The development of ideas, problem statements, and approaches to solutions by conceptualization, definition and design
2. Clarification	The formal expression of objectives, model structure, functional relationships, etc. through organization, documentation and accounting
3. Analysis	The interpretation and explanation of model and system behaviour through statistics, simulation and system analysis
4. Application	The use of the results of analysis and of models through communication, experimentation and technology transfer

Source: Adapted from Rykiel (1984).

Fig. 8.8. Sub-system definition using a circle diagram form of a conceptual model. In this instance the central focus is the profit generated by a sheep production system. (a) The central focus, profit, is split into income and expenditures or, more generally, input and output (1 and 2). The other numbers represent component inputs and outputs (or losses) with those numbers closest to the central focus representing inputs and outputs closest to the final grassland products. The arrows indicate connections among system components. Shading identifies sub-system components. (b) A representation of an extraction of one of the sub-systems within this model. Note that connections among system components make it possible to: (i) identify the sub-system and (ii) see how the sub-system affects the centre or objective. (c) An application of the model to investigate the effect on profit of the amount of herbage grown in the sheep production system. The sub-system of interest is indicated by shaded components and the connections among them. (From Spedding, 1975*a*.)

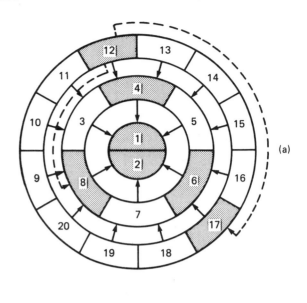

(a)

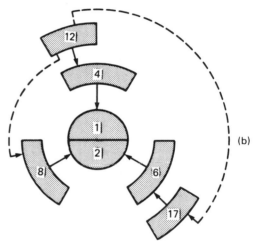

(b)

(Continued overleaf)

(c)

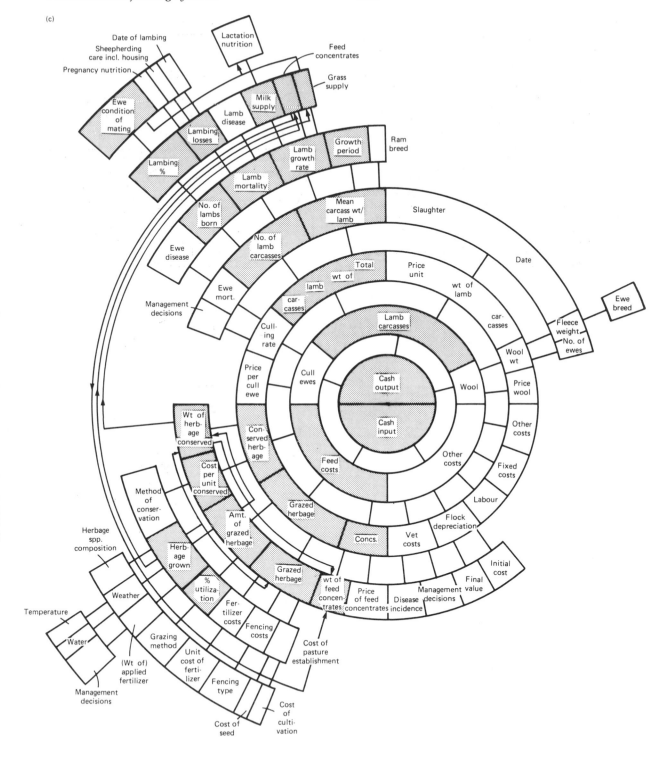

Fig. 8.8 (*continued*)

Fig. 8.9. A model of a national pasture plant domesticating system.

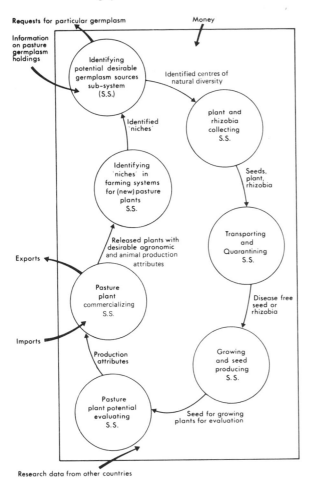

Requests for particular germplasm   Money

Information on pasture germplasm holdings

Identifying potential desirable germplasm sources sub-system (S.S.)

Identified centres of natural diversity

Identified niches

Identifying 'niches' in farming systems for (new) pasture plants S.S.

plant and rhizobia collecting S.S.

Seeds, plant, rhizobia

Released plants with desirable agronomic and animal production attributes

Exports

Pasture plant commercializing S.S.

Transporting and Quarantining S.S.

Disease free seed or rhizobia

Imports

Production attributes

Growing and seed producing S.S.

Pasture plant potential evaluating S.S.

Seed for growing plants for evaluation

Research data from other countries

Fig. 8.10. Forms of quantitative modelling which are used in the study of grassland systems. (From Wilson, 1984.)

	Steady state	Dynamic
Deterministic	Algebraic equations	Differential equations
Non-deterministic (stochastic)	Statistical and probability relationships	Discrete event simulation

Steady-state deterministic models include any algebraic relationship; they are well suited to problems concerned with the allocation of some limited resource (e.g. land, money, labour, fertilizer) where many alternatives exist. Linear programming is a well-known form of a steady state deterministic model. It involves numerical optimizing and has been frequently used by agricultural economists to study complex problems such as the minimization of costs or the maximization of profits, by animal nutritionists for determining the least cost or profit maximizing rations (e.g. Sauvant, Chapoutot & Lapierre, 1983) and more recently with the development of powerful algorithms for main-frame computers, for integrating biological and economic data from whole farming systems. Linear programming is also an appropriate tool for studying the integration of new grassland systems and cattle management options (e.g. Teitzel, Monypenny & Rogers, 1986). The technique of linear programming is described in Gass (1985).

Steady-state non-deterministic models apply where the mechanisms governing behaviour are not known, but where it can be assumed that certain variables are wholly or partially dependent on others. Seed yield, for instance, may be wholly or partly dependent on plant density (Fig. 4.4). Regression relates variables of interest. In its simplest form, e.g. seed yield (y) may be related linearly to plant density (x). The constants in linear regression, curvilinear regression, multi-linear regression and multiple regression are derived to minimize the sum of the errors squared at the discrete data points. Wilson (1984) includes probabilistic modelling in this section. Models with some element of randomness, to which we may assign a probability, are called *stochastic* models.

Steady-state modelling fits within Thornley's (1976) schema as empirical modelling. Empiricism involves attempting to fit some model or equation to data and then making deductions about the mechanisms involved. The opposite to this is mechanistic modelling, which attempts to understand the response of biological systems in terms of mechanisms. Mechanistic models are constructed by looking at the

8.11.2
Quantitative modelling

Quantification of processes within grassland systems aids both research and management. We have already used quantitative models extensively in this text, usually as algebraic equations or statistical relationships. As with all models these are a simplification of the real world and include assumptions made by the modeller; they are used mainly to predict the behaviour of some aspect of the real world and they require some form of validation. An understanding of this area of modelling may be gained from the schema proposed by Wilson (Fig. 8.10).

structure of the system and dividing it into its components in an attempt to understand the behaviour of the whole system in terms of the actions and interactions of its components (Thornley, 1976).

Dynamic, deterministic models include differential equations. Differential equations allow the simulation of situations in which time dependence can be represented continuously. State variables, i.e. variables that describe quantities or amounts (e.g. pasture dry matter, livestock weight, nitrogen content) are each associated with a rate variable which describes change with time (t), e.g. pasture growth rate, livestock growth rate, rate of nitrogen cycling. The value of a state variable at any point in time (T) can be derived by integration which, when repeated many times, completes the simulation of interest.

$$\text{RATE } (T - 1) = t \, [\text{STATE } (T - 1) \ldots] \quad (8.5)$$

$$\begin{aligned} \text{STATE } (T) = &[\text{RATE } (T - 1) \times DELT] \\ &+ \text{STATE } (T - 1) \quad (8.6) \end{aligned}$$

where RATE and STATE refer to any rate and state variable respectively, *DELT* is the time interval over which integration occurs the length of which is determined by the type of model.

Dynamic, non-deterministic models are concerned with simulation of discrete events using difference equations (as opposed to differential equations).

Quantitative models range over all levels of biological organization (Fig. 1.1). Morley & White (1985) describe quantitative models relevant to grassland agronomy ranging from ones concerned with nitrogen fixation at the cellular level to whole farm and national sectoral models. One useful division of models according to their use is between predictive, and explanatory or experimental models (P. Martin, 1986, personal communication):

(i) Predictive models. These are frequently concerned with grassland productivity and efficiency and thus with input/ouput relationships, the environment and production interactions and economics. This form of modelling contributes to estimations of the reliability of feed options, the answering of farm management questions such as 'how to use a given forage most efficiently with what type of animal?' and/or to estimations of the effect of variability in the weather and prices on enterprise performance. Predictive models also aid farm development decision-making by simulating development under variable conditions. The general structure of one such predictive model (McKeon & Rickert, 1986) begins with a water balance model, predicts monthly grassland growth and then considers a feed year, simulated on a seasonal time step (*DELT*, Eqn (8.6)).

(ii) Explanatory or experimental models. Predictive models can be run and re-run using e.g. historical climatic data and by varying the growth rate of the forage, the stocking rate, forage allocation and animal age classes (e.g. O'Leary, Connor & White, 1985). This allows experimentation on a whole farm basis and manipulation of variables which would be impracticable and too expensive to investigate with field research. From re-running a model with differing inputs or input–output relationships the researcher can predict how the output of the model might respond to varying factors. Hypotheses may be formulated, e.g. that fodder crop yield is dependent on the sowing date, and then tested by re-running the model.

8.11.3
Concluding remarks on grassland models

The modelling of grassland systems has adopted all of the approaches outlined above. Most grassland models have tackled specific aspects of grass or animal production; rarely have there been attempts to model the entire system mechanistically, even for a specific locality. One attempt at comprehensive modelling of grassland systems was the United States Grassland Biome study which generated the model described in Innis (1978).

All modelling is heuristic: it provides a learning experience which furthers investigation. Thus the researchers involved in modelling learn about the system or area they are attempting to model and where deficiencies in knowledge and understanding exist; these may be further explored by making predictions from the model or formulating hypotheses and testing them by conducting field or laboratory experiments or further modelling. Once constructed the models may also be used to help others learn, e.g. extension officers, farmers, other researchers. Such models need to be 'user-friendly'; one typical form is the 'what if?' model which, if formulated to, say, respond to user manipulation or farmers' questions about fertilizer application strategies, helps the farmer understand the dynamic interaction of fertilizer application rate, price, stocking rate, beef output and profit.

Problems relevant to agricultural modelling have been reviewed by Bennett & Macpherson (1985) and more specifically for grazing systems by Smith (1983), Freer & Christian (1983) and Emmans & Whittemore (1983). Authors conclude that modelling has yet to achieve its potential largely because of the lack of appreciation by modellers, biological researchers and administrators of many of the concepts and processes of modelling outlined above. We believe that future grassland agronomists will rely heavily on models, but it is worth concluding with the views of Bennett & Macpherson (1985) that the usefulness of the output from models is the key issue in modelling but that to date little has been done about checking that the outputs are useful and that they are worth the modelling effort.

8.12
Further reading
Conway, G. R. (1985). Agroecosystem analysis. *Agricultural Administration*, **20**, 31–55.

Jollans, J. L. (ed.) (1981). *Grassland in the British Economy.* Reading, UK: Centre for Agricultural Strategy.

Waterman, D. A. (1986). *A Guide to Expert Systems*. Reading, Massachusetts: Addison-Wesley.

Wilson, B. (1984). *Systems: Concepts, Methodologies and Applications.* Chichester, UK: John Wiley & Sons.

Derivation of climatic data for calculation of grassland growth

There are many formulae or models for calculating grassland growth, or tolerance of some processes, e.g. flowering, in relation to the environment. The most important environmental parameters are temperature, solar radiation and available water; the effects of these on growth were discussed briefly in Chapter 3.

Temperature is commonly measured or it can be interpolated between weather stations. Spatial variation in temperature is brought about mostly by altitude: temperatures fall by 0.65°C for every 100 m increase in altitude to 1.5 km (Lockwood, 1974).

Solar radiation is measured infrequently. Values for extra terrestrial radiation S (at the top of the atmosphere) can be found in Smithsonian Meteorological Tables (List, 1958). Alternatively they may be calculated as a function of latitude and day of the year. Thus S in MJ per m^2 per day is (Turner, 1979):

$$S = 117.06\,[(\cos L \times \cos d \times \sin N) + (N \times \sin L \times \sin d)]/(\pi z^2) \quad (A.1)$$

where L is the latitude (negative in the southern hemisphere), t is the day of the year (1 January being day 1 or $t = 1$), d is the angle in radians describing the seasonal sine-curve of solar radiation where:

$$d = 0.410\,15 \sin (0.017\,21t - 1.389) \quad (A.2)$$

N is the daylength and z is a term related to the earth–sun distance, in radians. N may be measured or alternatively, as a useful approximation, it may also be described as a function of latitude and day of the year:

$$N = 24 \cos^{-1} (-\tan L \times \tan d)/\pi \quad (A.3)$$

The term z can also be expressed in terms of the day of the year:

$$z = 1 + 0.016\,74(0.012\,71t - 1.622\,5) \quad (A.4)$$

The radiation, I, i.e. the amount of radiation which penetrates the atmosphere to the surface of the grassland, drives photosynthesis and evaporation. I can be calculated from S depending on the ratio of the number of hours of sunshine (n) to daylength, i.e. the number of hours between sunrise and sunset N:

$$I = (1 + 2n/N)(0.2S + 0.064) + 0.032 \quad (A.5)$$

This equation gives I in units of cal per cm^2 per min;

Fig. A.1. The relationship between net radiation I and total extraterrestrial radiation outside the earth's atmosphere S, showing that the relationship between I and S depends on the number of hours of sunshine n in a day of N h. (From Linacre, 1968.)

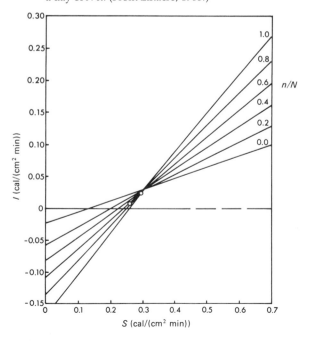

multiply by 0.041 8 to convert to MJ per m^2. The above expression assumes an albedo (vegetation reflectance) of 0.26 (Linacre, 1968). The relationship between I and S for varying ratios of sunshine hours to daylength (n/N) is shown in Fig. A.1.

Once I is obtained, we can calculate photosynthetic rates (see 'Further reading' in Chapter 3).

Available soil water is measured or determined by a water budget (Fig. A.2) for which there are now many computer routines (e.g. Taylor, Jordan & Sinclair, 1983). Briefly, the change in available soil water ΔAW, is given by:

$$\Delta AW = R + C + R_i - (R_0 + D + E) \qquad (A.6)$$

where R is the amount of rainfall, C is the amount of capillary rise from the subsoil to the root zone (usually ignored), R_i is the amount of run-on water, R_0 is the amount of run-off water, D is the amount of drainage water below the root zone and E is evaporation. Rainfall and evaporation are the most important terms in

Fig. A.2. Fate of rainfall: a schema to describe crop/soil water balance. (From Norman *et al.*, 1984.)

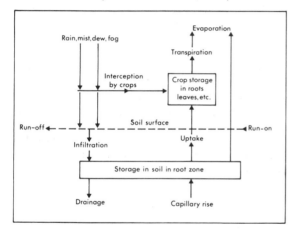

Eqn (A.6), at least on a regional (grassland system) scale.

Evaporation over long periods of time (so that there is virtually no storage or loss of heat by the crop) is given by:

$$E = I/\lambda \qquad (A.7)$$

where λ is the latent heat of evaporation of water: 2.453 MJ per kg at 20°C. Thus, using Eqn (A.7), if the average daily net radiation is 9 MJ per m^2, the average amount lost by evaporation is 9/2.453 or 4 mm per day.

The actual amount of water lost by evaporation from the vegetation plus the soil is approximated roughly from Eqn (A.7) but more accurately by measuring the net radiation absorbed by the vegetation, i.e. I above the grass minus I_s, the radiation reaching the soil surface. Furthermore, the actual (as distinct from the potential) amount of water lost by evaporation also depends on gradients of latent heat (temperature ΔT, °C) and water vapour (Δq, kg water) from the herbage to the atmosphere:

$$E = f[e(I - I_s)/\lambda + E_a]/(e + 1) \qquad (A.8)$$

where f is a constant which ranges from 0.6 to 0.8; e is the rate of change of vapour pressure per °C change in temperature at ambient temperature and E_a is a term used to describe the dependence of evaporation of u, the windspeed (km per day), at 2 m above the ground, and Δe, the vapour pressure deficit of the air (m bar):

$$E_a = 0.26(0.5 + 6.21 \times 10^{-3}u)(\Delta e) \qquad (A.9)$$

Eqn (A.8), which was proposed by H. L. Penman in 1948 (Penman, 1963) provides a unified concept of how evaporation relates quantitatively to radiation, temperature, humidity and windspeed. Eqn (A.8) or derivations usually underly budgets which calculate the amount of available soil water (Fig. 3.9) and water use and water use efficiency of grasslands (Table 3.1).

Calculating the feed demand of livestock

The liveweight of an animal is the major factor determining maintenance feed requirements. Extra feed is required for pregnancy, lactation and growth. The requirements of growing steers and heifers, as a function of liveweight and the rate at which liveweight is being gained per day, are given in Table B.1. In this worked example feed requirements related to a standard New Zealand pasture with an M/D value of 10.8, i.e. a metabolizable energy (ME) concentration of 10.8 MJ per kg dry matter (DM).

Methods for calculating the feed demand have been proposed by Milligan & McConnell (1976). They are:

(i) Adjust the DM requirement (using Tables, e.g. those in Milligan & McConnell (1976) and MAFF (1975)) by the relative ME value of the feed. For example, the DM requirement of a 300-kg steer growing at 0.5 kg per day is 5.7 kg (Table B.1). If

hay having a relative ME value of 0.78 is fed, then:

5.7/0.78 = 7.3 kg hay would be required.

(ii) Divide the ME requirements by the ME concentration of the feed. For example, a 300-kg steer growing at 0.5 kg per day requires 62 MJ ME per day. The ME concentration of good pasture hay is 8.4. Therefore the DM needed is:

62/8.4 or 7.3 kg hay.

Feed demand values for single animals may be extrapolated to calculate the total feed demand of a herd. This involves determining the average liveweight in the herd and multiplying by the number of animals and the number of days in the budget period. A worked example for a theoretical dairy herd is provided in Table B.2. Alderman (1983) reviews the development of feeding systems based on the ME.

Table B.1. *Daily feed requirements of growing steers and heifers*

| Liveweight (kg) | Liveweight gain/day (kg) | | | | | | | | | | | | |
| | 0 | | 0.25 | | 0.50 | | 0.75 | | 1.00 | | 1.25 | | 1.50 | |
	MJ ME	kg DM[a]	MJ ME	kg DM[a]	MJ ME	kg DM[a]	MJ ME	kg DM[a]	MJ ME	kg DM[a]	MJ ME	kg DM[a]	MJ ME	kg DM[a]
100	24	2.2	28	2.6	33	3.1	37	3.4	42	3.9	47	4.4	53	4.9
150	30	2.8	35	3.2	41	3.8	47	4.4	54	5.0	60	5.6	68	6.3
200	34	3.1	41	3.8	48	4.4	56	5.2	64	5.9	73	6.8	82	7.6
250	39	3.6	47	4.4	55	5.1	65	6.0	74	6.9	84	7.8	95	8.8
300	43	4.0	52	4.8	62	5.7	73	6.8	83	7.7	95	8.9	107	9.9
350	47	4.4	58	5.4	69	6.4	80	7.4	93	8.8	106	9.8	119	11.0
400	51	4.7	63	5.8	75	6.9	88	8.1	101	9.4	116	10.7	131	12.1
450	55	5.1	67	6.2	81	7.5	85	7.9	110	10.2	125	11.6	142	13.1
500	59	5.5	72	6.7	87	8.1	102	9.4	118	10.9	135	12.5	153	14.2

[a] Assuming a mixed-length leafy pasture with an ME concentration of 10.8 MJ/kg DM and a relative ME value of 1.00. For pastures of different ME concentration the requirement by the animal of DM (kg) = MJ ME required/(M/D of pasture).
Source: Milligan & McConnell (1976).

Table B.2. *Animal requirements of a theoretical dairy herd*

Assume the average mass of cows in the herd to be 550 kg; then the total animal requirements in terms of metabolizable energy (ME) and digestible crude protein (DCP) are (MAFF, 1976; R. G. Packham, personal communication, 1986):

Maintenance
Daily allowance for maintenance = 59 MJ ME and 325 g DCP
Add 5% activity allowance to energy requirement, ME requirement = 62 MJ
Assume no overall weight gain or loss over the whole herd, i.e. those cows losing weight are balanced by those gaining weight

Production
Assume the average butterfat of milk is running at about 3.6% and the solids-not-fats (SNF) at 8.6%
The desirable figures are 4.0% butterfat and 8.8% SNF
Nutrient allowances for this are 5.3 MJ ME/kg milk and 53 g DCP/kg milk
A pregnancy allowance should be made for cows during the last 4 months of pregnancy, or from 8 months post-calving:

Maintenance + pregnancy allowance = 82 MJ ME and 465 g DCP

The average herd production of milk is about 9700 l/week or 1386 l/day, produced from 82 milkers and 20 dry cows; this equals 16.9 l/day/cow in milk

The specific gravity of milk is approximately 1.03, therefore the weight of milk produced = 1427 kg/day

Assuming a calving interval of 365 days average for the herd, then:
  the ME requirements in MJ/day are:
    Maintenance + pregnancy for 34 cows = 34 × 82 =   2788
    Maintenance for 68 cows = 68 × 325   =   4216
    Lactation allowance = 1427 × 5.3   =   7563
  the total ME required/day   = 14567 MJ

  the DCP requirements in kg DCP/day are:
    Maintenance + pregnancy for 34 cows = 34 × 465/1000 =   15.81
    Maintenance for 68 cows = 68 × 325/1000   =   22.10
    Lactation allowance = 1427 × 5.3   =   75.63
and the total DCP required/day   = 113.54

APPENDIX C

Investment analysis using discounted cash flow

The first step in discounted cash flow analysis is to determine a physical plan and then to estimate relevant costs and returns. These costs and returns are then compared on a 'with and without basis' as in, say, 'with a new pasture compared to without'. This proposed change can be seen as a project. This project is then separated from existing activities on the farm. Project cash flows are the net cash flow of the farm without the new project minus the net cash flow with the project; they may be divided into three classes:

(i) project income or benefits (b), i.e. a cash inflow, as e.g. from increased sales of livestock due to increased stocking rates;

(ii) annual operating costs (c), i.e. a cash outflow;

(iii) capital costs (k), i.e. a cash outflow, usually in the early years of the project.

In any year $t (= 1, 2, 3$, etc.) the net cash flow (a_t) is thus:

$$a_t = b_t - (c_t + k_t) \qquad (C.1)$$

Some problems are encountered in allocating costs such as the farmer's labour, where e.g. the proposed project may take more time, and opportunity costs, where the project may influence the demand or use on some other component of the farming system. Also, estimates should be made of the economic life of the project. This will be influenced by the life of the most durable asset, and is often 10 years or less; with the case of land or agroforestry it will be longer. When a cut-off point is decided the depreciated value of the assets needs to be included as a benefit in that particular year.

Tax and potential tax savings and interest rates must be predicted. These are value-laden judgements (Madden, 1985) and because of the long-term nature of pasture development projects this form of analysis necessitates comparison of cash flows at widely removed points in time. Cash flows are thus dis-counted to present-day values or compounded (as in compound interest) to some common point of time in the future.

The rate of interest chosen is central to this analysis. The rate of interest chosen may be based on the borrowing rate; the interest rate paid on loan finance; an opportunity cost, i.e. the interest rate that converted funds could earn in the most financially attractive alternative use including off-farm investments; and a subjective rate of time preference decided upon by the analyst's view of the future.

Present value

In a project a sequence of amounts of money $a_0, a_1, a_2 \ldots a_n$ is earned or lost at the end of years (t) 0, 1, 2, ... Discounting converts these future amounts of money to their present value (PV):

$$PV_n = a_0 - \frac{a_1}{1 + i} + \frac{a_2}{(1 + i)^2} + \cdots + \frac{a_n}{(1 + i)^n} \qquad (C.2)$$

$$= \sum_{t=0}^{n} \frac{a_t}{(1 + i)^t} \qquad (C.3)$$

where i is the rate of compound interest expressed as a decimal. The present values of all expected cash inflows (B_n) and cash outflows (D_n) arising from a project such as pasture improvement can be calculated:

$$B_n = \sum_{t=0}^{n} \frac{b_t}{(1 + i)^t} \qquad (C.4)$$

and

$$D_n = \sum_{t=0}^{n} \frac{d_t}{(1 + i)^t} \qquad (C.5)$$

where b_t and d_t are the expected cash inflow and

outflow in year t and i is again the compound rate of interest. Thus the net present value (NPV) of a project, which is its net contribution to a farmer's wealth, is the difference between Eqns (C.4) and (C.5):

$$NPV = B_n - D_n \qquad (C.6)$$

The NPV may be calculated using annual net cash flows, i.e. costs and returns do not have to be specified separately. Thus Eqn (C.3) can substitute for Eqn (C.6) where $a_t = b_t - (c_t + k_t)$ (Eqn (C.1)).

An investment resulting in a positive NPV is normally considered acceptable but this depends on such factors as:

(i) the NPV of other possible projects, a higher NPV being the more profitable;

(ii) estimates of the risk involved, which may be partly gauged by carrying out a sensitivity analysis;

(iii) the payback period, i.e. the minimum number of years required for the project to recover its costs;

(iv) the benefit to costs ratio (B/C), which is used to rank investments with different magnitudes of capital outlay:

$$\text{Gross B/C ratio} = \frac{PV_b}{PV_k + PV_c} \qquad (C.7)$$

(v) the internal rate of return (IRR), i.e. that rate of interest for which the net present value of the cash flow series is zero, i.e. the value of i for which

$$\sum_{t=0}^{n} \frac{a_t}{(1 + r)^t} = 0 \qquad (C.8)$$

or alternatively, the IRR may be recorded as the interest rate at which the PV of cash inflows equals the PV of cash outflows;

(vi) the rate of inflation built into the analysis;

(vii) the availability of credit.

In summary a viable investment will have NPV $\geq$ 0, a B/C ratio ≥ 1 and an IRR $\geq$ the opportunity cost of investing money or borrowing money. Risk and uncertainty must be considered; e.g. a sensitivity analysis may be carried out on the above criteria and payback period.

A worked example for establishing an improved permanent phalaris (*Phalaris aquatica*) pasture in arable country degraded by weed invasion in the tablelands of south-east Australia is provided in Table C.1. The analysis is for 10 ha and is based on local prices and costs, i.e. the dollar values should be ignored as these are site-specific, and the example is provided only to demonstrate how the analysis might be conducted.

Table C.1. *Discounted cash flow analysis of benefits of establishing permanent phalaris pasture on arable country in tablelands of south-east Australia*

(a) *Information about the farm: the increase in stocking rate expected after establishing permanent pasture*[a]

Enterprise unit (ha)	10.0
Wool price ($(US)/kg)	4.3
Wool cut (kg)	6.5
Stocking rate	d.s.e./ha[b]
Year 1	2.0
Year 2	6.5
Year 3	8.5
Years 4–10	12.0

(b) *Stocking schedule: potential number of sheep which could be grazed on the improved pasture and, therefore, the need for animal purchases following pasture improvement*

Years	1	2	3	4	5	6	7	...	10
Potential	20	65	85	120	120	120	120		120
On hand s/y[c]	0	17	55	72	102	102	102		102
Purchases	20	48	30	48	18	18	18		18
Culls	2	7	9	12	12	12	12		12
Deaths	1	3	4	6	6	6	6		6
On hand e/y[c]	17	55	72	102	102	102	102		102

(c) Costs of pasture establishment, returns from extra livestock and, therefore, the cash balance (cumulative debit or credit) following pasture improvement

		Years 1	2	3	4	5	6	7	8	9	10
Costs ($(US)/ha)											
Plough	@ 17.50	175									
Scarify twice	@ 8.00	160									
Rabbit control	@ 2.00	20									
Sow	@ 15.00	150									
Seed	@ 34.60	346									
Seed treatment	@ 1.35	14									
Single super-phosphate	@ 164.25/t	205		205		205		205		205	
Superphosphate application	@ 31.10/t	39		39		39		39		39	
Allowance for failure											111
Stock purchase	@ 22.00/head	440	1 056	655	1 051	396	396	396	396	396	396
Stock costs	@ 7.50/head	150	488	638	900	900	900	900	900	900	900
Total costs		1 699	1 544	1 537	1 951	1 540	1 296	1 540	1 296	1 540	1 407
Returns ($(US)/ha)											
Wool sales		559	1 817	2 376	3 354	3 354	3 354	3 354	3 354	3 354	3 354
c.f.a. sales[d]	@ 11.00/head	22	72	94	132	132	132	132	132	132	132
Salv. value	@ 22.00/head										2 244
Total returns		581	1 889	2 470	3 486	3 486	3 486	3 486	3 486	3 486	5 730
Net returns		−1 118	345	933	1 536	1 946	2 190	1 946	2 190	1 946	4 323
Interest		−168	−141	−22	0	0	0	0	0	0	0
Cumulative balance		−1 285	−1 082	−171	1 365	3 310	5 300	7 446	9 636	11 582	15 905

NPV	6 628
Interest on borrowed funds	0.15
Discount rate	0.15

[a] This example refers to 10 ha fields in a 550–700 mm annual average rainfall zone having high soil fertility and previously growing subterranean clover and nitrophilous weeds.
[b] d.s.e., dry sheep equivalents: one castrated male sheep = 1 d.s.e., a beef breeder = 12.00 d.s.e.; based on body weight.
[c] s/y, start of year; e/y, end of year.
[d] c.f.a. sales; cast for age sales, i.e. sheep sold at the end of their productive life.
Source: Adapted from Muir (1985).

REFERENCES

Ackoff, R. L. (1962). *Scientific Method: Optimizing Applied Research Decisions*. 464 pp. New York: John Wiley & Sons.

Agricultural Research Council (1980). *The Nutrition Requirements of Livestock*. Slough, UK: Commonwealth Agricultural Bureaux.

Aitken, Y. (1974). *Flowering Time, Climate and Genotype*. 193 pp. Melbourne: Melbourne University Press.

Aitken, Y. (1985). Temperate herbage grasses and legumes. In *Handbook of Flowering*, Vol. 1, ed. A. H. Halevy, pp. 185–202. Boca Raton, Florida: CRC Press.

Akpan, E. E. J. & Bean, E. W. (1977). The effects of temperature upon seed development in three species of forage grasses. *Annals of Botany*, **41**, 689–95.

Alderman, G. (1983). Status of the new European feeding systems. In *Feed Information and Animal Production*, ed. G. E. Robards & R. G. Packham, pp. 157–74. Slough, UK: Commonwealth Agricultural Bureaux.

Alexander, R. T. (1985). Effect of sheep grazing regime on performance of mature prairie grass. *Proceedings of the New Zealand Grassland Association*, **46**, 151–6.

Ali, T. & Stobbs, T. H. (1980). Solubility of the protein of tropical pasture species and the rate of its digestion in the rumen. *Animal Feed Science and Technology*, **5**, 183–92.

Allden, W. G. (1982). Cattle growth in a mediterranean environment. *Australian Meat Research Committee Review*, **42**, 1–20.

Allden, W. G. & Whittaker, I. A. McD. (1970). Determinants of herbage intake by grazing sheep: the interrelationship of factors influencing herbage intake and availability. *Australian Journal of Agricultural Research*, **21**, 755–66.

Allen, H. P. (1979). Renewing pastures by direct drilling. In *Changes in Sward Composition and Productivity*, ed. A. H. Charles & R. J. Haggar, pp. 217–22. Reading: British Grassland Society.

Allen, P. G. (1984). A control program for spotted alfalfa aphid in dryland lucerne pasture and its implementation in South Australia. *Proceedings of the 4th Australian Applied Entomological Research Conference, Adelaide*, pp. 114–20. Adelaide, Australia.

Alli, I., Robidas, E., Noroozi, E. & Baker, B. E. (1985). Some changes associated with the field drying of lucerne and timothy. *Grass and Forage Science*, **40**, 221–6.

Anderson, J. P. E., Armstrong, R. A. & Smith, S. N. (1981). Methods to evaluate pesticide damage to the biomass of the soil microflora. *Soil Biology and Biochemistry*, **13**, 149–53.

Andrew, C. S. & Kamprath, E. J. (eds.) (1978). *Mineral Nutrition of Legumes in Tropical and Subtropical Soils*. Melbourne: Commonwealth Scientific and Industrial Research Organization.

Andrew, M. H. & Mott, J. J. (1983). Annuals with transient seed banks: the population biology of indigenous *Sorghum* species of tropical north west Australia. *Australian Journal of Ecology*, **8**, 265–76.

Andrews, P., Collins, W. J. & Stern, W. R. (1977). The effect of withholding water during flowering on seed production in *Trifolium subterraneum* L. *Australian Journal of Agricultural Research*, **28**, 301–7.

Anslow, R. C. & Green, J. O. (1967). The seasonal growth of pasture grasses. *Journal of Agricultural Science*, **68**, 109–22.

Argel, P. J. & Humphreys, L. R. (1983). Environmental effects on seed development and hardseededness in *Stylosanthes hamata* cv. Verano I. Temperature. *Australian Journal of Agricultural Research*, **34**, 261–70.

Armstrong, D. G. (1982). Digestion and utilization of energy. In *Nutritional Limits to Animal Production from Pastures*, ed. J. B. Hacker, pp. 225–44. Slough, UK: Commonwealth Agricultural Bureaux.

Arndt, W. (1965). The impedance of soil seals and the forces of emerging seedlings. *Australian Journal of Soil Research*, **3**, 55–68.

Arnold, G. W. (1964). Factors within plant associations affecting the behaviour and performance of grazing animals. In *Grazing in Terrestrial and Marine Environments*, ed. D. J. Crisp, pp. 133–54. Oxford: Blackwell.

Bacon, C. W., Lyons, P. C., Porter, J. K. & Robbins, J. D. (1986). Ergot toxicity from endophyte-infected grasses: a review. *Agronomy Journal*, **78**, 106–16.

Bailey, R. W. (1973). Structural carbohydrates. In *Chemistry and Biochemistry of Herbage*, Vol. I, ed. G. W. Butler & R. W. Bailey, pp. 157–211. London: Academic Press.

Baker, C. K. & Gallagher, J. N. (1983). The development of winter wheat in the field 1. *Journal of Agricultural Science*, **101**, 327–35.

Baldwin, R. L. (1984). Digestion and metabolism of ruminants. *Bio Science*, **34**, 244–9.

Ball, P. R. & Field, T. R. O. (1985). Productivity and economics of legume-based pastures and grass swards receiving fertilizer nitrogen in New Zealand. In *Forage Legumes for Energy-Efficient Animal Production*, ed. R. F. Barnes, P. R. Ball, R. W. Brough, G. C. Martin & D. J.

Minson, pp. 47–55. Springfield, Virginia: US Department of Agriculture Agricultural Research Service.

Ball, P. R. & Keeney, D. R. (1983). Nitrogen losses from urine-affected areas of a New Zealand pasture, under contrasting seasonal conditions. *Proceedings of the XIVth International Grassland Congress*, pp. 342–4. Boulder, Colorado: Westview Press.

Barlow, N. D. (1985). Laboratory studies on porina (*Wiseana* spp.) feeding behaviour: the 'functional response' to herbage mass. *Proceedings of the 4th Australasian Conference on the Ecology of Grassland Invertebrates*. Christchurch, New Zealand.

Barnard, C. (ed.) (1966). *Grasses and Grasslands*. London: Macmillan.

Barnard, C. & Frankel, O. H. (1966). Grass, grazing animals and man in historic perspective. In *Grasses and Grasslands*, ed. C. Barnard, pp. 1–12. London: Macmillan.

Barrow, N. J. (1975). The response to phosphate of two annual pasture species II. *Australian Journal of Agricultural Research*, **26**, 145–56.

Basuno, E. & Petheram, R. J. (1985). A village profile for livestock component FSR in Java. *Farming Systems Support Project, Networking Paper No. 4*. Gainesville, Florida: FSSP.

Bawden, R. J. (1985). Problem-based learning: An Australian Perspective. In *Problem Based Learning in Education for the Professions*, ed. D. Boud, pp. 43–57. Sydney: Higher Education Research and Development Society of Australia.

Bawden, R. J., Ison, R. L., Macadam, R. D., Packham, R. G. & Valentine, I. (1985). A research paradigm for systems agriculture. In *Agricultural Systems Research for Developing Countries*, ed. J. V. Remenyi, pp. 31–42. Canberra: Australian Centre for International Agricultural Research.

Beebe, J. (1985). Rapid rural appraisal: the critical first step in a farming systems approach to research. *Farming Systems Support Project Networking Paper No. 5*. Gainesville, Florida: FSSP.

Bekendam, J. & Grob, R. (1979). *Handbook for Seedling Evaluation*. Zurich, Switzerland: International Seed Testing Association.

Bell, M. K. (1974). Decomposition of herbaceous litter. In *Biology of Plant Litter Decomposition, Vol. 1*, eds. C. H. Dickinson & G. J. F. Pugh, pp. 37–68. London: Academic Press.

Belyuchenko, I. S. (1977). Features of growth of paniculated and eragrostoid perennial crops. *Proceedings of the 13th International Grassland Congress*, **1**, 43. Berlin: Akademie-Verlag.

Bement, R. E. (1968). Plains pricklypear: relations to grazing intensity and blue grama yield on the central Great Plains. *Journal of Range Management*, **21**, 83–6.

Bennett, D. & Macpherson, D. K. (1985). Structuring a successful modelling activity. In *Agricultural Systems Research for Developing Countries*, ed. J. V. Remenyi, pp. 70–6. Canberra: Australian Centre for International Agricultural Research.

Bergersen, F. J. (1982). *Root Nodules of Legumes: Structure and Functions*. Chichester: Research Studies Press, John Wiley.

Bircham, J. S. (1984). Patterns of herbage growth during lactation and level of herbage mass at lambing: their significance to animal production. *Proceedings of the New Zealand Grassland Association*, **45**, 177–83.

Bircham, J. S. & Hodgson, J. (1983). The influence of sward condition on rates of herbage growth and senescence in mixed swards under continuous stocking management. *Grass and Forage Science*, **38**, 323–32.

Birks, P. R. & Allen, P. G. (1969). Pasture cockchafer. *Journal of the Department of Agriculture, South Australia*, **73**, 39–43.

Bishop, H. G., Walker, B. & Rutherford, M. T. (1983). Renovation of tropical legume grass pastures in northern Australia. *Proceedings of the XIVth International Grassland Congress*, pp. 555–8. Boulder, Colorado: Westview Press.

Bjorkman, O., Boynton, J. & Berry, J. (1976). Comparison of the heat stability of photosynthesis, chloroplast membrane reactions, photosynthetic enzymes, and soluble protein in leaves of heat-adapted and cold-adapted C4 species. *Carnegie Institute of Washington Yearbook*, **75**, 400–7.

Black, J. W. & Kenney, P. A. (1984). Factors affecting diet selection by sheep II. Height and density of pasture. *Australian Journal of Agricultural Research*, **35**, 565–78.

Black, J. W., Kenney, P. A. & Colebrook, W. F. (1987). Diet selection by sheep. In *Temperate Pastures: their Production, Use and Management*, ed. J. L. Wheeler, C. J. Pearson & G. E. Robards, pp. 331–4. Melbourne: Australian Wool Corporation and Commonwealth Scientific and Industrial Research Organization.

Blaser, R. E. (1982). Integrated pasture and animal management. *Tropical Grasslands*, **16**, 9–24.

Boertje, R. D. (1985). An energy model for adult female caribou of the Donali herd, Alaska. *Journal of Range Mangement*, **38**, 468–73.

Bolan, N. S., Robson, A. D. & Barrow, N. J. (1983). Plant and soil factors including mycorrhizal infection causing sigmoidal response of plants to applied phosphorus. *Plant and Soil*, **73**, 187–201.

Bolsen, K. K. (1986). New technology in forage conservation-feeding systems. *Proceedings of the XVth International Grassland Congress*, pp. 82–8. Nishi-nasuno, Japan: The Japanese Society of Grassland Science.

Bowen, H. D. (1981). Alleviating mechanical impedance. In *Modifying the Root Environment to Reduce Crop Stress*, ed. G. F. Arkin & H. M. Taylor, pp. 21–60. St Joseph, Michigan: American Society of Agricultural Engineers.

Box, E. (1978). Geographical dimensions of terrestrial net and gross primary productivity. *Radiation and Environmental Biophysics*, **15**, 305–22.

Breman, H. & Wit, C. T. de (1983). Rangeland productivity and exploitation in the Sahel. *Science*, **221**, 1341–7.

Brewbaker, J. (1986). Leguminous trees and shrubs for S.E. Asia and the South Pacific. In *Forages in South East Asia and the South Pacific*, ed. G. J. Blair, D. A. Ivory & T. R. Evans, pp. 43–50. Canberra: Australian Centre for International Agricultural Research.

Bromfield, S. M., Cumming, R. W., David, D. J. & Williams, C. H. (1983). Change in soil pH, manganese and aluminium under subterranean clover pasture. *Australian Journal of Experimental Agriculture and Animal Husbandry*, **23**, 181–91.

Brooking, I. R. (1979). Temperature effects on inflorescence development and grain filling of cereals. *Proceedings of the Agronomy Society of New Zealand*, **9**, 65–9.

Brougham, R. W. (1959). The effects of season and weather on the growth rate of ryegrass and clover pasture. *New Zealand Journal of Agricultural Research*, **2**, 283–96.

Brouwer, R. (1962). Distribution of dry matter in the plant. *Netherlands Journal of Agricultural Science*, **10**, 361–76.

Brown, K. R. & Evans, R. S. (1973). Animal treading – a review of the work of the late D. B. Edmond. *New Zealand Journal of Experimental Agriculture*, **1**, 217–26.

Bryan, W. W. & Evans, T. R. (1973). Effects of soils, fertilizers and stocking rates on pastures and beef production on the Wallum of South-Eastern Queensland. 1. Botanical composition and chemical effects on plants and soils. *Australian Journal of Experimental Agriculture and Animal Husbandry*, **13**, 516–29.

Buckley, R. C. (ed.) (1982*a*). *Ant-Plant Interactions in Australia.* 162 pp. The Hague: Dr. W. Junk.

Buckley, R. C. (1982*b*) Ant-plant interactions: a world review. In *Ant-Plant Interactions in Australia*, ed. R. C. Buckley, pp. 111–41. The Hague: Dr. W. Junk.

Bula, R. J., Lechtenberg, V. L., Hold, D. A., Humphreys, L. R., Crowder, L. V. & Box, T. W. (1977). *Potential of the world's forages for ruminant animal production.* 90 pp. Morrilton, Arkansas: Winrock International Livestock Research and Training Centre.

Burdon, J. J. (1978). Mechanisms of disease control in heterogeneous plant populations – an ecologist's view. In *Plant Disease Epidemiology*, ed. P. R. Scott & A. Bainbridge, pp. 193–200. Oxford: Blackwell.

Burdon, J. J. (1983). *Trifolium repens. Journal of Ecology*, **71**, 307–30.

Burns, I. G. (1980). A simple model for predicting the effects of winter leaching of residual nitrate on the nitrogen fertilizer need for spring crops. *Journal of Soil Science*, **31**, 187–202.

Burns, J. C. & Standaert, J. E. (1985). Productivity and economics of legume-based vs. nitrogen-fertilized grass-based pastures in the United States. In *Forage Legumes for Energy-Efficient Animal Production*, ed. R. F. Barnes, P. R. Ball, R. W. Brough, G. C. Martin & D. J. Minson, pp. 56–71. Springfield, Virginia: US Department of Agriculture Agricultural Research Service.

Butler, G. W. (1973). Mineral biochemistry of herbage. In *Chemistry and Biochemistry of Herbage*, ed. G. W. Butler & R. W. Bailey, Vol. 2, pp. 127–62. London: Academic Press.

Butler, G. W. & Bailey, R. W. (eds.) (1973). *Chemistry and Biochemistry of Herbage*, Vol. I. 639 pp. London: Academic Press.

Cable, D. R. & Tschirley, F. H. (1961). Response of native and introduced grasses following aerial spraying of velvet mesquite in southern Arizona. *Journal of Range Management*, **14**, 155–9.

Cameron, D. F. (1967). Flowering in Townsville lucerne (*Stylosanthes humilis*). 2. The effect of latitude and time of sowing on the flowering time of single plants. *Australian Journal of Experimental Agriculture and Animal Husbandry*, **7**, 495–500.

Campbell, M. H. (1979). Aerial seeding in Iran. *Agricultural Gazette of N.S.W.*, **90**, 6–8.

Campbell, M. H. (1982). Restricting losses of aerially sown seed due to seed-harvesting ants. In *Ant-Plant Interactions in Australia*, ed. pp. R. C. Buckley, 25–30. The Hague: Dr. W. Junk.

Campbell, M. H. & Swain, F. C. (1973*a*). Effect of strength, tilth and heterogeneity of the soil surface on the radicle-entry of surface-sown seeds. *Journal of the British Grassland Society*, **26**, 355–9.

Campbell, M. H. & Swain, F. G. (1973*b*). Factors causing losses during the establishment of surface-sown pastures. *Journal of Range Management*, **26**, 355–9.

Catchpoole, V. R., Harper, L. A. & Myers, R. J. K. (1983). Annual losses of ammonia from a grazed pasture fertilized with urea. *Proceedings of the XIVth International Grasslands Congress*, pp. 344–7. Boulder, Colorado: Westview Press.

Cavers, P. B. (1963). Comparative biology of *Rumex obtusifolius* L. and *R. crispus* L. including the variety triganulatus. PhD thesis, University of Wales. Cited by Harper (1977).

Chapman, D. F. (1983). Growth and demography of *Trifolium repens* stolons in grazed hill pastures. *Journal of Applied Ecology*, **20**, 597–608.

Charles-Edwards, D. A. (1982). *Physiological Determinants of Crop Growth.* 161 pp. Sydney: Academic Press.

Charles-Edwards, D. A., Cockshull, K. E., Horridge, J. S. & Thornley, J. H. M. (1979). A model of flowering in *Chrysanthemum. Annals of Botany*, **44**, 557–66.

Checkland, P. B. (1981). *Systems Thinking, Systems Practice.* 330 pp. Chichester, UK: John Wiley & Sons.

Checkland, P. B. (1985). Reflections on 'Towards a systems-based methodology for real-world problem solving'. In *Agricultural Systems Education*, ed. K. Wilson. Honolulu, Hawaii: University of Hawaii.

Chestnutt, D. M. B. (1984). Effect of weight change on the performance of autumn-calving suckler cows. *Journal of Agricultural Science*, **102**, 353–9.

Christian, K. R. (1987). Matching pasture production and animal requirements. In *Temperate Pastures: their Use, Production and Management*, ed. J. L. Wheeler, C. J. Pearson & G. E. Robards, pp. 463–76. Melbourne: Australian Wool Corporation and Commonwealth Scientific and Industrial Research Organization.

Christie, E. K. (1984). Natural Grasslands. In *Control of Crop Productivity*, ed. C. J. Pearson, pp. 199–218. Sydney: Academic Press.

Christie, E. K. & Detling, J. K. (1982). Analysis of interference between C3 and C4 North American grasses in relation to temperature and soil nitrogen supply. *Ecology*, **63**, 1277–84.

CIAT (Centro Internacional de Agricultura Tropical) (1980). *Annual Report Tropical Pastures Programme 1979.* 156 pp. Cali, Colombia: CIAT.

CIAT (Centro Internacional de Agricultura Tropical). (1983). *Annual Report Tropical Pastures Programme.* 375 pp. Cali, Colombia: CIAT.

Clarkson, D. T. (1981). Nutrient interception and transport by root systems. In *Physiological Processes Limiting Plant Productivity*, ed. C. B. Johnson, pp. 307–30. London: Butterworth.

Clarkson, N. M. & Russell, J. S. (1975). Flowering responses to vernalisation and photoperiod in annual medics (*Medicago* spp.). *Australian Journal of Agricultural Research*, **26**, 831–8.

Clarkson, N. M. & Russell, J. S. (1976). Effect of water stress on the phasic development of annual *Medicago* species. *Australian Journal of Agricultural Research*, **27**, 227–34.

Clarkson, N. M. & Russell, J. S. (1979). Effect of temperature on the development of two annual medics. *Australian Journal of Agricultural Research*, **30**, 909–16.

Clawson, M. (1969). Systems analysis of natural resources and crop production. In *Physiological Aspects of Crop Yield*, ed. J. D. Eastin, F. A. Haskins, C. Y. Sullivan & C. H. M. van Bavel, pp. 1–14. Madison: American Society of Agronomy.

Clements, R. J., Hayward, M. D. & Byth, D. E. (1983). Genetic adaptation in pasture plants. In *Genetic Resources of Forage Plants*, ed. J. G. McIvor & R. A. Bray, pp. 101–16. Melbourne: Commonwealth Scientific and Industrial Research Organization.

Clements, R. J. & Ludlow, M. M. (1977). Frost avoidance and frost resistance in *Centrosema virginianum. Journal of Applied Ecology*, **14**, 551–66.

Clewett, J. F., McCown, R. L. & Leslie, J. K. (1986). Tropical pastures in the farming system: integration of tropical crops and pastures. In *Proceedings of the 3rd Australian Conference on Tropical Pastures*, ed. G. J. Murtagh & R. M.

Jones, pp. 110–18. Brisbane: Tropical Grassland Society of Australia.

Cochrane, T. T. (1975) Land use classification in the lowlands of Bolivia. In *Soil Management in Tropical America*, pp. 109–25. Raleigh, North Carolina: North Carolina State University.

Coe, M. J., Cumming, D. H. & Phillipson, J. (1976). Biomass and production of large African herbivores in relation to rainfall and primary production. *Oecologia*, **22**, 341–54.

Colbourn, P. & Dowdell, R. J. (1984). Denitrification in field soils. *Plant and Soil*, **76**, 213–26.

Collins, W. J. (1978). The effect of defoliation on inflorescence production, seed yield and hardseededness in swards of subterranean clover. *Australian Journal of Agricultural Research*, **29**, 789–801.

Collins, W. J. (1981). The effects of length of growing season, with and without defoliation, on seed yield and hardseededness in swards of subterranean clover. *Australian Journal of Agricultural Research*, **32**, 783–92.

Collins, W. J. & Aitken, Y. (1970). The effect of leaf removal on flowering time in subterranean clover. *Australian Journal of Agricultural Research*, **21**, 893–903.

Collins, W. J., Rhodes, J., Rossiter, R. C. & Palmer, M. J. (1983). The effect of defoliation on seed yield of two strains of subterranean clover grown in monocultures and in binary mixtures. *Australian Journal of Agricultural Research*, **34**, 671–9.

Collins, W. J., Rossiter, R. C. & Ramos Monreal, A. (1978). The influence of shading on seed yield in subterranean clover. *Australian Journal of Agricultural Research*, **29**, 1167–75.

Collis-George, N. & Melville, M. D. (1975). Water absorption by swelling seeds. I. Constant surface boundary conditions. *Australian Journal of Soil Research*, **13**, 141–58.

Collis-George, N. & Melville, M. D. (1978). Water absorption by swelling seeds. II. Surface condensation boundary conditions. *Australian Journal of Soil Research*, **16**, 291–310.

Conway, G. R. (1985a). Agricultural ecology and farming systems research. In *Agricultural Systems Research for Developing Countries*, ed. J. V. Remenyi, pp. 43–59. Canberra: Australian Centre for International Agricultural Research.

Conway, G. R. (1985b). Agroecosystem analysis. *Agricultural Administration*, **20**, 31–55.

Cook, S. J. (1980). Establishing pasture species in existing swards: a review. *Tropical Grasslands*, **14**, 181–7.

Cook, S. J., Blair, G. J. & Lazenby, A. (1978). Pasture degeneration II. The importance of superphosphate, nitrogen and grazing management. *Australian Journal of Agricultural Research*, **29**, 19–29.

Cooper, C. S. (1977). Growth of the legume seedling. *Advances in Agronomy*, **29**, 119–39.

Cooper, J. P. (1963). Species and population differences in climatic responses. In *The Environmental Control of Plant Growth*, ed. L. T. Evans, pp. 381–400. New York: Academic Press.

Cooper, J. P. (1964). Climatic variation in forage grasses 1. Leaf development in climatic races of *Lolium* and *Dactylis*. *Journal of Applied Ecology*, **1**, 81–90.

Cooper, J. P. & McWilliam, J. R. (1966). Climatic variation in forage grasses. II. *Journal of Applied Ecology*, **3**, 191–212.

Corbett, J. L. (1987). Energy and protein utilization by grazing animals. In *Temperate Pastures: their Production, Use and Management*, ed. J. L. Wheeler, C. J. Pearson & G. E. Robards, pp. 341–55. Melbourne: Australian Wool Corporation and Commonwealth Scientific and Industrial Research Organization.

Cornish, P. S. (1987). Root growth and function in temperate pastures. In *Temperate Pastures: their Production, Use and Management*, ed. J. L. Wheeler, C. J. Pearson & G. E. Robards, pp. 79–98. Melbourne: Australian Wool Corporation and Commonwealth Scientific and Industrial Research Organization.

Corrall, A. J. (1978). The effect of genotype and water supply on the seasonal pattern of grass production. *Proceedings of the 7th General Meeting of the European Grassland Federation.* Section 2. pp. 23–32. Gent, Belgium.

Cox, G. W. (1984). The linkage of inputs to outputs in agroecosystems. In *Agricultural Ecosystems. Unifying Concepts*, ed. R. Lowrance, B. R. Stinner & G. J. House, pp. 187–208. New York: John Wiley & Sons.

Crawley, M. J. (1983). *Herbivory: The Dynamics of Animal–Plant Interactions*. Studies in Ecology, Vol. 10. London, UK: Blackwell.

Crofts, F. C., Geddes, H. J. & Carter, O. G. (1963). *Water Harvesting and Planned Pasture Production at Badgery's Creek.* 33 pp. Sydney: University of Sydney, School of Agriculture Report No. 6.

Crush, J. R. (1974). Plant responses to vesicular arbuscular mycorrhizas VII. *New Phytologist*, **73**, 743–9.

CSIRO (Commonwealth Scientific and Industrial Research Organization) (1975). Replacing oestrogenic clovers. *Rural Research*, **87**, 12–18.

CSIRO (Commonwealth Scientific and Industrial Research Organization) (1979). How some tropical legumes survive frost. *Rural Research*, **99**, 27–8.

Culvenor, C. C. (1970). Toxic plants – a re-evaluation. *Search*, **1**, 103–10.

Culvenor, C. C. (1973). Alkaloids. In *Chemistry and Biochemistry of Herbage*, ed. G. W. Butler & R. W. Bailey, Vol. 1, pp. 375–446. London: Academic Press.

Culvenor, C. C. (1987). Detrimental factors in pastures and forages. In *Temperate Pastures: their Production, Use and Management*, ed. J. L. Wheeler, C. J. Pearson & G. E. Robards, pp. 435–45. Melbourne: Australian Wool Corporation and Commonwealth Scientific and Industrial Research Organization.

Curll, M. L. & Davidson, J. L. (1983). Defoliation and productivity of a *Phalaris* – subterranean clover sward, and the influence of grazing experience on sheep intake. *Grass and Forage Science*, **38**, 159–67.

Curll, M. L. & Wilkins, R. J. (1983). The comparative effects of defoliation, treading and excreta on a *Lolium perenne* – *Trifolium repens* pasture grazed by sheep. *Journal of Agricultural Science*, **100**, 451–60.

Curll, M. L. & Wilkins, R. J. (1985). The effect of cutting for conservation on a grazed perennial ryegrass – white clover pasture. *Grass and Forage Science*, **40**, 19–30.

Curll, M. L., Wilkins, R. J., Snaydon, R. W. & Shanmugalingam, V. S. (1985). The effects of stocking rate and nitrogen fertilizer on a perennial ryegrass – white clover sward. 1. Sward and sheep performance. *Grass and Forage Science*, **40**, 129–40.

Curry, J. P. & Bolger, T. (1984). Growth, reproduction and litter and soil consumption by *Lumbricus terrestris* L. in reclaimed peat. *Soil Biology and Biochemistry*, **16**, 253–7.

Dale, J. E. & Milthorpe, F. L. (1983a). General features of the production and growth of leaves. In *The Growth and Functioning of Leaves*, ed. J. E. Dale & F. L. Milthorpe, pp. 151–77. Cambridge University Press.

Dale, J. E. & Milthorpe, F. L. (eds.) (1983*b*). *The Growth and Functioning of Leaves*. 540 pp. Cambridge University Press.

Dann, P. R. & Coombe, J. B. (1987). Utilization of fodder crops and crop residues. In *Temperate Pastures: their Production, Use and Management*, ed. J. L. Wheeler, C. J. Pearson & G. E. Robards, pp. 517–55. Melbourne: Australian Wool Corporation and Commonwealth Scientific and Industrial Research Organization.

Date, R. A., Batthyany, C. & Jaureche, C. (1965). Survival of rhizobia in inoculated and pelleted seed. *Proceedings of the IXth International Grassland Congress*, Vol. 1, pp. 263–9. Sao Paulo: Canposto a empresso ebicoes alaido limitida.

Davies, A. (1971). Growth rates and crop morphology in vernalised and non vernalised swards of perennial ryegrass in spring. *Journal of Agricultural Science*, **77**, 273–82.

Davies, A. & Thomas, H. (1983). Rates of leaf and tiller production in young spaced perennial ryegrass plants in relation to soil temperature and solar radiation. *Annals of Botany*, **57**, 591–7.

Davies, W. I. C. & Davies, J. (1981). Varying the time of spraying with paraquat or glyphosate before direct drilling of grass and clover seeds with and without calcium peroxide. *Grass and Forage Science*, **36**, 65–9.

Deboer, J. (1983). Design of alternative crop-livestock systems. *Proceedings of the Crop–Livestock Research Workshop, Asian Cropping Systems Network*, pp. 38–59. Los-Banos, Philippines: International Rice Research Institute.

Decker, A. M. & Dudley, R. F. (1976). Minimum tillage establishment of five forage species using five sod-seeding units and two herbicides. In *Hill Lands*, ed. J. Luchok, J. D. Cawthon & M. J. Breslin, pp. 140–5. Morgantown, West Virginia, USA: West Virginia University Books.

Deinum, B. & Dirven, J. G. P. (1976). Climate, nitrogen and grass 7. Comparisons of production and chemical composition of *Brachiaria ruziziensis* and *Setaria sphacelata* grown at different temperatures. *Netherlands Journal of Agricultural Science*, **24**, 67–78.

Demarquilly, C. & Weiss, Ph. (1970). *Tabeux de valeur alimentaires des fourrages*. 65 pp. Versailles: Ministere de l'Agriculture Institute National de la Recherche Agronomique.

Denmead, O. T., Freney, J. R. & Simpson, J. R. (1976). A closed ammonia cycle within a plant canopy. *Soil Biology and Biochemistry*, **8**, 161–4.

Dent, J. B. & Blackie, M. J. (1979). *Systems Simulation in Agriculture*. 180 pp. London: Applied Science.

Deregibus, V. A., Sanchez, R. A. & Casal, J. J. (1983). Effects of light quality on tiller production in *Lolium* spp. *Plant Physiology*, **72**, 900–2.

Devitt, A. C., Quinlivan, B. J. & Francis, C. M. (1978). The flowering of annual legume pasture species and cultivars in Western Australia. *Australian Journal of Experimental Agriculture and Animal Husbandry*, **18**, 75–80.

Dickenson, C. H. & Pugh, G. J. F. (eds.) (1974). *Biology of Plant Litter Decomposition*, Vol. 1. 146 pp. London: Academic Press.

Dijk, G. van & Hoogervorst, N. (1983). The demand for grasslands in Europe towards 2000. Some implications of a possible scenario. In *Efficient Grassland Farming*, ed. A. J. Corrall, pp. 21–31. Hurley, UK: British Grassland Society.

Doak, B. W. (1952). Some chemical changes in the nitrogenous constituents of urine when voided on pasture. *Journal of Agricultural Science*, **31**, 162–71.

Donald, C. M. (1963). Competition among crop and pasture plants. *Advances in Agronomy*, **15**, 1–114.

Douglas, G. K. (1985). The meanings of agricultural sustainability. In *Agricultural Sustainability in a Changing World Order*, ed. G. K. Douglas, pp. 3–29. Boulder, Colorado: Westview Press.

Dowling, P. M. & Linscott, D. L. (1983). Use of pesticides to determine relative importance of pest and disease factors limiting establishment of sod-seeded lucerne. *Grass and Forage Science*, **38**, 179–85.

Doyle, C., Corrall, A. J., Du, Y., le. & Thomas, C. (1982). Integration of conservation with grazing. In *Milk from Grass*, ed. C. Thomas & J. W. O. Young, pp. 59–74. Hurley, UK: ICI Agricultural Division and Grassland Research Institute.

Doyle, C. J. & Elliot, J. G. (1983). Putting an economic value on increases in grass production. *Grass and Forage Science*, **38**, 169–77.

Doyle, C. J. & Lazenby, A. (1984). The effect of stocking rate and fertilizer usage on income variability for dairy farms in England and Wales. *Grass and Forage Science*, **39**, 117–27.

Du, Y. L. P., le., Combellas, J., Hodgson, J. & Baker, R. D. (1979). Herbage intake and milk production by grazing dairy cows 2. The effects of level of winter feeding and daily herbage allowance. *Grass and Forage Science*, **34**, 249–60.

Dudar, Y. A. & Machado, R. (1981). Seed maturation and dispersal of grasses in Cuba. *Pastos y Forrages*, **4**, 175–200.

Dudzinski, M. L. & Arnold, G. W. (1973). Comparisons of diets of sheep and cattle grazing together on sown pastures on the Southern tablelands of New South Wales by principal components analysis. *Australian Journal of Agricultural Research*, **24**, 899–912.

Dure, L. S. (1975). Seed formation. *Annual Review of Plant Physiology*, **26**, 259–78.

Dyne, G. M., van & Heady, H. F. (1965). Botanical composition of sheep and cattle diets on a mature annual range. *Hilgardia*, **36**, 465–92.

Edmond, D. B. (1958). The influence of treading on pasture. A preliminary study. *New Zealand Journal of Agricultural Research*, **1**, 319–28.

Edye, L. A. & Grof, B. (1983). Selecting cultivars from naturally occurring genotypes: evaluating *Stylosanthes* species. In *Genetic Resources of Forage Plants*, ed. J. G. McIvor & R. A. Bray, pp. 217–36. Melbourne: Commonwealth Scientific and Industrial Research Organization.

Edye, L. A., Williams, W. T. & Winter, W. H. (1978). Seasonal relations between animal gain, pasture production and stocking rate on two tropical grass–legume pastures. *Australian Journal of Agricultural Science*, **29**, 103–11.

Egan, A. R., Wanapat, M., Doyle, P. T., Dixon, R. M. & Pearce, G. R. (1986). Animal production limitations imposed by intake, digestibility and rate of passage. In *Forages in South East Asian and South Pacific Agriculture*, ed. G. J. Blair, D. A. Ivory & T. R. Evans, pp. 104–10. Canberra: Australian Centre for International Agricultural Research.

Ehara, K. & Tanaka, S. (1961). Effects of temperature on the growth behaviour and chemical composition of the warm and cool season grasses. *Proceedings of the Crop Science Society of Japan*, **29**, 304–6.

Ehara, K., Yamada, Y. & Maeno, N. (1966). The effect of defoliation on the carbon balance in *Dactylis glomerata*. *Annals of Botany*, **30**, 185–98.

Elton, C. S. (1930). *Animal Ecology and Evolution*. Oxford.

Emmans, G. C. & Whittemore, C. T. (1983). Simulation models in nutritional management of livestock. In *Feed Information and Animal Production*, ed. G. E. Robards & R. G. Packham, pp. 323–32. Slough, UK: Commonwealth Agricultural Bureaux.

Evans, L. T. (1964). Reproduction. In *Grasses and Grasslands*, ed. C. Barnard, pp. 126–53. London: Macmillan.

Evans, L. T. (1969). *The Induction of Flowering.* 488 pp. Sydney: Macmillan.

Evans, R. A. & Young, J. A. (1970). Plant litter and the establishment of alien annual weed species in rangeland communities. *Weed Science*, **18**, 697–703.

Everist, S. L. (1981). *Poisonous Plants of Australia.* London: Angus and Robertson.

FAO (United Nations Food and Agriculture Organization) (1979). *1978 Production Yearbook*, Vol. 32. Rome: FAO.

Felippe, G. M. (1979). The flowering of tillers of *Panicum maximum.* Jacq. *Revta Brasil Botany*, **2**, 87–90.

Fick, G. W. (1984). Physiological responses of plants to pest damage. In *Insect Pest Management Modelling*, ed. C. A. Shoemaker & W. G. Ruesink. New York: Wiley-Interscience.

Field, T.R.O. (1980). Can we manipulate the annual pattern of pasture growth? *Proceedings of the New Zealand Grassland Society*, **41**, 80–8.

Field, T. R. O., Pearson, C. J. & Hunt, L. A. (1976). Effects of temperature on the growth and development of alfalfa (*Medicago sativa* L.). *Herbage Abstracts*, **46**, 145–50.

Fisher, M. J., Charles-Edwards, D. A. & Campbell, N. A. (1980). A physiological approach to the analysis of crop growth data. 2. Growth of *Stylosanthes humilis. Annals of Botany*, **46**, 425–34.

Fitzpatrick, E. A. & Nix, H. A. (1970). The climate factor in Australian grassland ecology. In *Australian Grasslands*, ed. R. M. Moore, pp. 3–26. Canberra: Australian National University Press.

Fletcher, G. M. & Dale, J. E. (1974). Growth of tiller buds in barley: effects of shade treatment and mineral nutrition. *Annals of Botany*, **38**, 63–76.

Flood, R. G. & Halloran, G. M. (1982). Flowering behaviour of four annual grass species in relation to temperature and photoperiod. *Annals of Botany*, **49**, 469–75.

Forbes, D. K. & Tribe, D. E. (1970). The utilization of roughages by sheep and kangaroos. *Australian Journal of Zoology*, **18**, 247–56.

Forbes, T. D. A. & Hodgson, J. (1985). Comparative studies of the influence of sward conditions on the ingestive behaviour of cows and sheep. *Grass and Forage Science*, **40**, 69–77.

Ford, C. W., Morrison, I. M. & Wilson, J. R. (1979). Temperature effects on lignin, hemicellulose and cellulose in tropical and temperate grasses. *Australian Journal of Agricultural Research* **30**, 621–33.

Fowkes, N. D. & Landsberg, J. J. (1981). Optimal root systems in terms of water uptake and movement. In *Mathematics and Plant Physiology*, ed. D. A. Rose & D. A. Charles-Edwards, pp. 109–28. London: Academic Press.

Frame, J. (1981). Herbage mass. In *Sward Measurement Handbook*, ed. J. Hodgson, R. D. Baker, A. Davies, A. S. Laidlaw & J. D. Leaver, pp. 39–70. Hurley, UK: British Grassland Society.

Francis, C. M. & Devitt, A. C. (1969). The effect of waterlogging on the growth and isoflavone concentration of *Trifolium subterraneum* L. *Australian Journal of Agricultural Research*, **20**, 819–25.

Francis, C. M. & Gladstones, J. S. (1974). Relationships among rate and duration of flowering and seed yield components in subterranean clover (*Trifolium subterraneum*). *Australian Journal of Agricultural Research*, **25**, 435–42.

Frederick, L. R. (1978). Effectiveness of rhizobia–legume association. In *Mineral Nutrition of Legumes in Tropical and Subtropical Soils*, ed. C. S. Andrew & E. J. Kamprath, pp. 265–76. Melbourne: Commonwealth Scientific and Industrial Research Organization.

Freer, M. & Christian, K. R. (1983). Application of feeding standard systems to grazing ruminants. In *Feed Information and Animal Production*, ed. G. E. Robards & R. G. Packman, pp. 333–56. Slough, UK: Commonwealth Agricultural Bureaux.

Fukai, S. & Silsbury, J. H. (1976). Responses of subterranean clover communities to temperature I. *Australian Journal of Plant Physiology*, **3**, 527–43.

Fussell, L. K. & Pearson, C. J. (1980). Effects of grain development and thermal history on grain maturation and seed vigour of *Pennisetum americanum. Journal of Experimental Botany*, **31**, 635–43.

Fussell, L. K., Pearson, C. J. & Norman, M. J. T. (1980). Effect of temperature during various growth stages on grain development and yield of *Pennisetum americanum. Journal of Experimental Botany*, **31**, 621–34.

Gallagher, R. T., White, E. P. & Mortimer, P. H. (1981). Ryegrass staggers: isolation of potent neurotoxins lolitrem A and lolitrem B from staggers-producing pasture. *New Zealand Veterinary Journal*, **29**, 189–90.

Garcia-Huidobro, J., Monteith, J. L. & Squire, G. R. (1982a). Time, temperature and germination of pearl millet (*Pennisetum typhoides* S. & H.). I. Constant temperature. *Journal of Experimental Botany*, **33**, 288–96.

Garcia-Huidobro, J., Monteith, J. L. & Squire, G. R. (1982b). Time, temperature and germination of pearl millet (*Pennisetum typhoides* S. & H.). II. Alternating temperature. *Journal of Experimental Botany*, **33**, 297–302.

Gardener, C. J. (1980). Diet selection and liveweight performance of steers on *Stylosanthes hamata* – native grass pastures. *Australian Journal of Agricultural Research*, **31**, 379–92.

Gardener, C. J. (1981). Population dynamics and stability of *Stylosanthes hamata* cv. Verano in grazed pastures. *Australian Journal of Agricultural Research*, **33**, 63–74.

Garwood, E. A. & Sinclair, J. (1979). The use of water by six grass species 2. Root distribution and use of soil water. *Journal of Agricultural Science*, **93**, 25–35.

Gass, S. I. (1985). *Linear Programming: Methods and Applications.* 532 pp. New York: McGraw-Hill.

Gifford, G. F. & Hawkins, R. H. (1978). Hydrologic impact of grazing on infiltration: a critical review. *Water Resources Research*, **14**, 305–13.

Gill, A. M. (1981). Adaptive responses of Australian vascular plant species to fires. In *Fires and the Australian Biota*, ed. A. M. Gill, R. H. Groves & I. R. Noble. 582 pp. Canberra: Australian Academy of Science.

Gillard, P. (1979). Improvements of native pasture with Townsville stylo in the dry tropics of sub-coastal northern Queensland. *Australian Journal of Experimental Agriculture and Animal Husbandry*, **19**, 325–36.

Gillard, P. (1982). Beef cattle production from improved pastures. *World Animal Review*, **44**, 2–8.

Gordon, A. J., Ryle, G. J. A., Mitchell, D. F. & Powell, C. E. (1985). The flux of ^{14}C-labelled photosynthate through soybean root nodules during N_2 fixation. *Journal of Experimental Botany*, **36**, 756–69.

Gordon, F. J. (1980). Grass and fresh grass products. *Proceedings of the Nutrition Society*, **39**, 249–56.

Goss, R. L. (1972). Nutrient uptake and assimilation for

quality turf versus maximum vegetative growth. In *The Biology and Utilization of Grasses*, ed. V. B. Younger & C. M. McKell, pp. 278–91. New York: Academic Press.

Graham, N. McC. (1983). The energy value of livestock feeds: alternative expressions and their usefulness as feed attributes. In *Feed Information and Animal Production* ed. G. E. Robards & R. G. Packham, pp. 157–74. Slough, UK: Commonwealth Agricultural Bureaux.

Gramshaw, D. & Stern, W. R. (1977*a*). Survival of annual ryegrass (*Lolium rigidum* Gaud.) in a Mediterranean environment I. Effect of summer grazing by sheep on seed numbers and seed germination in autumn. *Australian Journal of Agricultural Research*, **28**, 81–91.

Gramshaw, D. & Stern, W. R. (1977*b*). Survival of annual ryegrass (*Lolium rigidum* Gaud.) seed in a Mediterranean type environment. II. Effects of short-term burial on persistence of viable seed. *Australian Journal of Agricultural Research*, **28**, 93–101.

Graswami, A. K. & Willcox, J. S. (1969). Effects of applying increasing levels of nitrogen to ryegrass I. Composition of various nitrogenous fractions and free amino acids. *Journal of the Science of Food and Agriculture*, **20**, 592–9.

Greenwood, D. J. (1981). Crop response to agronomic practice. In *Mathematics and Plant Physiology*, ed. D. A. Rose & D. A. Charles-Edwards, pp. 195–216. London: Academic Press.

Greenwood, D. J., Clever, T. J., Turner, M. K., Hunt, J., Niendorf, K. B. & Loquens, S. M. H. (1980). Comparison of the effects of potassium fertilizer on the yield, potassium content and quality of 22 different vegetable and agricultural crops. *Journal of Agricultural Science*, **95**, 441–85.

Greenwood, D. J., Wood, J. T., Cleaver, T. J. & Hunt, J. (1971). A theory for fertilizer response. *Journal of Agricultural Science*, **77**, 511–23.

Gregory, F. G. & Sen, P. K. (1937). Physiological studies in plant nutrition IV. The relation of respiration rate to the carbohydrate and nitrogen metabolism of the barley leaf as determined by N and K deficiency. *Annals of Botany*, **1**, 521–61.

Grime, J. P. (1979). *Plant Strategies and Vegetation Processes*. 222 pp. Chichester: Wiley.

Grossbard, E. & Atkinson, D. (eds.) (1985). *The Herbicide Glyphosate*. 490 pp. London: Butterworth.

Gruen, F. H. (1959). Pasture improvement – the farmer's economic choice. *Australian Journal of Agricultural Economics*, **3**, 19–44.

Gutteridge, R. C. (1983). Productivity of forage legumes on rice paddy walls in Northeast Thailand. *Proceedings of the XIVth International Grassland Congress*, pp. 226–9. Boulder, Colorado: Westview Press.

Hacker, J. B. (ed.) (1982). *Nutritional Limits to Animal Production from Pastures*. Farnham Royal: Commonwealth Agricultural Bureaux.

Hacker, J. B. (1984). Genetic variation in seed dormancy in *Digitaria milanjiana* in relation to rainfall at the collection site. *Journal of Applied Ecology*, **21**, 947–59.

Hacker, J. B., Andrew, M. H., McIvor, J. G. & Mott, J. J. (1984). Evaluation in contrasting climates of dormancy characteristics of seed of *Digitaria milanjiana*. *Journal of Applied Ecology*, **21**, 961–70.

Haggar, R. J. & Bastian, C. J. (1980). Regulating the content of white clover in mixed swards using grass-suppressing herbicides. *Grass and Forage Science*, **35**, 129–37.

Hamilton, B. A., Hutchinson, K. J., Annis, P. C. & Donnelly, J. B. (1973). Relationships between the diet selected by grazing sheep and the herbage on offer. *Australian Journal of Agricultural Research*, **24**, 271–7.

Hamilton, W. T. & Scrifes, C. J. (1982). Prescribed burning during winter for maintenance of buffelgrass. *Journal of Range Management*, **35**, 9–12.

Hampton, J. C. (1984). Control measures for ergot in the paspalum (*Paspalum dilatatum* Poir.) seed crop. *Journal of Applied Seed Production*, **2**, 32–5.

Hampton, J. G. & Hebblethwaite, P. D. (1983). Yield components of the perennial ryegrass (*Lolium perenne* L.) seed crop. *Journal of Applied Seed Production*, **1**, 23–5.

Hardarson, G., Heichel, G. H. & Barnes, D. K. (1982). Rhizobial strain preference of alfalfa populations selected for characteristics associated with N_2 fixation. *Crop Science*, **22**, 55–8.

Hardarson, G., Heichel, G. H., Vance, C. P. & Barnes, D. K. (1981). Evaluation of alfalfa and *Rhizobium meliloti* for compatibility in nodulation and nodule effectiveness. *Crop Science*, **21**, 562–7.

Haresign, W. (1981). Body condition, milk yield and reproduction in cattle. In *Recent Developments in Ruminant Nutrition*, ed. W. Haresign & D. J. A. Cole, pp. 1–16. London: Butterworth.

Harlan, J. R. (1983). The scope for collection and improvement of forage plants. In *Genetic Resources of Forage Plants*, ed. J. G. McIvor & R. A. Bray, pp. 3–16. Melbourne: Commonwealth Scientific and Industrial Research Organization.

Harper, J. L. (1977). *Population Biology of Plants*. 892 pp. London: Academic Press.

Harper, J. L. & McNaughton, I. H. (1962). The comparative biology of closely related species living in the same area. *New Phytologist*, **61**, 175–88.

Harris, C. E. & Tullberg. J. N. (1980). Pathways of water loss from legumes and grasses cut for conservation. *Grass and Forage Science*, **35**, 1–11.

Harris, W. (1974). Competition among pasture plants V. Effects of frequency and height of cutting on competition between *Agrostis tenuis* and *Trifolium repens*. *New Zealand Journal of Agricultural Research*, **17**, 251–6.

Hart, R. H. (1972). Forage yield, stocking rate and beef gains on pasture. *Herbage Abstracts*, **42**, 345–53.

Haveren, B. P. van (1983). Soil bulk density as influenced by grazing intensity and soil type on a short grass prairie site. *Journal of Range Management*, **36**, 586–8.

Haynes, R. J. (1983). Soil acidification induced by leguminous crops. *Grass and Forage Science*, **38**, 1–11.

Heady, H. F. (1986). Grazing as a part of agroforestry. *Proceedings of the XVth International Grassland Congress*, pp. 1070–1. Nishi-nasuno, Japan: The Japanese Society of Grassland Science.

Hebblethwaite, P. D. (ed.) (1980). *Seed Production*. 694 pp. London and Boston: Butterworth.

Hebblethwaite, P. D., Wright, D. & Noble, A. (1980). Some physiological aspects of seed yield in *Lolium perenne* L. (perennial ryegrass). In *Seed Production*, ed. P. D. Hebblethwaite, pp. 71–90. London: Butterworth.

Hecke, P., van, Impens, I. & Behaeghe, T. J. (1981). Temporal variation of species composition and species diversity in permanent grassland plots with different fertilizer treatment. *Vegetatio*, **47**, 221–32.

Heichel, G. H. (1985). Energy budgets for legume-based vs. nitrogen-fertilized grass-based pastures in the United States. In *Forage Legumes for Energy-Efficient Animal Production*, ed. R. F. Barnes, P. R. Ball, R. W. Brougham, G.

C. Martin & D. J. Minson, pp. 72–80. Springfield, Virginia: US Department of Agriculture.

Heichel, G. H., Barnes, D. K. & Vance, C. P. (1981). Nitrogen fixation of alfalfa in the seeding year. *Crop Science*, **21**, 330–5.

Heitschmidt, R. K., Price, D. L., Gordon, R. A. & Frasure, J. R. (1982). Short duration grazing at the Texas experimental ranch: effects on aboveground net primary production and seasonal growth dynamics. *Journal of Range Management*, **35**, 367–72.

Hennessy, J. T., Gibbens, R. P., Tromble, J. M. & Cardenas, M. (1983). Vegetation changes from 1935 to 1980 in mesquite dunelands and former grasslands of southern new Mexico. *Journal of Range Management*, **36**, 370–4.

Henzell, E. F. (1983). Contribution of forages to worldwide food production: now and in the future. *Proceedings of the XIVth International Grassland Congress*, pp. 42–7. Boulder, Colorado: Westview Press.

Herbel, C. H., Morton, H. L. & Gibbens, R. P. (1985). Controlling shrubs in the arid southwest with tebuthiuron. *Journal of Range Management*, **38**, 391–4.

Heydecker, W. (1978). 'Primed' seeds for better crop establishment? *Span*, **21**, 12–14.

Hight, G. K., Sinclair, D. P. & Lancaster, R. J. (1968). Some effects of shading and of nitrogen fertilizer on the chemical composition of freeze-dried and oven-dried herbage, and on the nutritive value of oven-dried herbage fed to sheep. *New Zealand Journal of Agricultural Research*, **11**, 286–302.

Hilder, E. J. (1964). The distribution of plant nutrients by sheep at pasture. *Proceedings of the Australian Society of Animal Production*, **5**, 241–8.

Hill, M. J. (1980). Temperate pasture grass-seed crops: formative factors. In *Seed Production*, ed. P. D. Hebblethwaite, pp. 137–49. London: Butterworth.

Hill, M. J. & Pearson, C. J. (1985). Primary growth and regrowth responses of temperate grasses to different temperatures and cutting frequencies. *Australian Journal of Agricultural Research*, **36**, 25–34.

Hill, M. J., Pearson, C. J. & Kirby, A. C. (1985). Germination and seedling growth of prairie grass, tall fescue and Italian ryegrass at different temperatures. *Australian Journal of Agricultural Research*, **36**, 13–24.

Hochman, Z., Godyn, D. L. & Scott, B. J. (1987). The integration of data on lime use by modelling. In *Soil Acidity and Plant Growth*, ed. A. D. Robson, W. M. Porter & J. S. Yeates. Sydney: Academic Press.

Hocking, P. J., Steer, B. T. & Pearson, C. J. (1984). Nitrogen nutrition of non-leguminous crops. A review. *Field Crops Abstract*, **37**, 625–36; **37**, 721–41.

Hodgkinson, K. C. (1974). Influence of partial defoliation on photosynthesis, photo respiration and transpiration by lucerne leaves of different ages. *Australian Journal of Plant Physiology*, **1**, 561–78.

Hodgkinson, K. C., Harrington, G. N., Griffin, G. F., Noble, J. C. & Young, M. D. (1984). Management of vegetation with fire. In: *Management of Australia's Rangelands*, ed. G. N. Harrington, A. D. Wilson & M. D. Young, pp. 141–56. Melbourne: Commonwealth Scientific and Industrial Research Organization.

Hodgkinson, K. C. & Quinn, J. A. (1978). Environmental and genetic control of reproduction in *Danthonia caespitosa* populations. *Australian Journal of Botany*, **26**, 351–64.

Hodgson, H. C. (1964). The protein and amino-acid composition and nitrogen distribution in two tropical grasses. *Journal of the Science of Food and Agriculture*, **15**, 721–4.

Hodgson, J. (1981). The influence of variations in the surface characteristics of the sward upon the short-term rate of herbage intake by calves and lambs. *Grass and Forage Science*, **36**, 49–57.

Hodgson, J. (1986). The significance of sward characteristics in the management of temperate sown pastures. *Proceedings of the XVth International Grassland Congress*, pp. 63–7. Nishi-nasuno, Japan: The Japanese Society of Grassland Science.

Hodgson, J., Capriles, J. M. R. & Fenlon, J. S. (1977). The influence of sward characteristics on the herbage intake of grazing calves. *Journal of Agricultural Science*, **89**, 743–50.

Hogan, J. P., Kenney, P. A. & Weston, R. H. (1987). Factors affecting the intake of feed by grazing animals. In *Temperate Pastures: their Production, Use and Management*, ed. J. W. Wheeler, C. J. Pearson & G. E. Robards, pp. 317–27. Melbourne: Australian Wool Corporation and Commonwealth Scientific and Industrial Research Organization.

Hollingsworth, D. F. (1981). The influence of nutrition on the future demand for grassland products. In *Grassland in the British Economy*, ed. J. L. Jollans, pp. 118–34. Reading, UK: Centre for Agricultural Strategy.

Holmes, W. (ed.) (1980). *Grass, its Production and Utilization.* 295 pp. Oxford: Blackwell Scientific.

Howe, C. D. & Chancellor, R. J. (1983). Factors affecting the viable seed content of soils beneath lowland pastures. *Journal of Applied Ecology*, **20**, 915–22.

Hubbard, C. F. (1944). Taxonomy, description and distribution of species and varieties. In *Imperata cylindrica*, pp. 5–12. Imperial Agricultural Bureaux Publication 7. Hurley, England.

Humphreys, L. R. (1979). *Tropical Pasture Seed Production.* 143 pp. FAO Plant Production and Protection Paper 8. Rome: United Nations Food and Agriculture Organization.

Humphreys, L. R. (1981). *Environmental Adaptation of Tropical Pasture Plants.* 261 pp. London: Macmillan.

Humphreys, L. R. (1984). Grazing management and the persistence of yield in tropical pasture legumes. *Asian Pastures*, Food and Fertilizer Technology Centre Book Series No. 25, pp. 1–11.

Hunt, L. A. (1965). Some implications of death and decay in pasture production. *Journal of the British Grassland Society*, **20**, 27–31.

Hur, S. N. & Nelson, C. J. (1985). Cotyledon and leaf development associated with seedling vigor of six forage legumes. *Proceedings of the XVth International Grassland Congress*, pp. 374–6. Nishi-nasuno, Japan: The Japanese Society of Grassland Science.

Hyde, E. O. C. (1954). The function of the hilum in some *Papilionaceae* in relation to the ripening of the seed and the permeability of the testa. *Annals of Botany*, **18**, 241–56.

Ibrahim, M. N. M. (1983). In *The Utilization of Fibrous Agricultural Residues*, ed. G. R. Pearce. 53 pp. Canberra: Australian Government Publications.

Innis, G. S. (ed.) (1978). *Grassland Simulation Model.* 298 pp. New York: Springer-Verlag.

Inosaka, M., Ito, K., Numaguchi, H. & Misumi, M. (1978). Studies on the productivity of some tropical grasses VI. *Japanese Journal of Tropical Agriculture*, **21**, 71–6.

IRRI (International Rice Research Institute) (1983). *Crop–Livestock Research Workshop.* 406 pp. Los Banos, Philippines.

Ison, R. L. & Hopkinson, J. M. (1985). Pasture grasses and legumes of warm climate regions. In *Handbook of Flowering*, Vol 1, ed. A. H. Halevy, pp. 208–52. Boca Raton, Florida: CRC Press.

Ison, R. L. & Humphreys, L. R. (1984*a*). Flowering of *Stylosanthes guianensis* in relation to juvenility and the long–short day requirement. *Journal of Experimental Botany*, **35**, 121–6.

Ison, R. L. & Humphreys, L. R. (1984*b*). Effects of temperature on the flowering and seed production of *Stylosanthes guianensis* cultivars. *Annals of Applied Biology*, **104**, 347–55.

Ison, R. L. & Humphreys, L. R. (1984*c*). Reproductive physiology of *Stylosanthes*. In *The Biology and Agronomy of Stylosanthes*, ed. H. M. Stace & L. A. Edye, pp. 257–77. Sydney: Academic Press.

Israel, D. W. & Jackson, W. A. (1978). The influence of nitrogen nutrition on ion uptake and translocation by leguminous plants. In *Mineral Nutrition of Legumes in Tropical and Subtropical Soils*, ed. C. S. Andrew & E. K. Kamprath, pp. 113–29. Melbourne: Commonwealth Scientific and Industrial Research Organization.

ISTA (International Seed Testing Association) (1976). International rules for seed testing – rules, 1976. *Seed Science and Technology*, **4**, 3–49.

Jaakkola, A. (1984). Leaching losses of nitrogen from a clay soil under grass and cereal crops in Finland. *Plant and Soil*, **76**, 59–66.

Jackson, D. L. & Jacobs, S. W. L. (1985). *Australian Agricultural Botany*. 377 pp. Sydney: University of Sydney Press.

Jacobs, V. E. & Stricker, J. A. (1976). Economic comparisons of legume nitrogen and fertilizer nitrogen in pastures. In *Biological N Fixation in Forage–Livestock Systems*, ed. C. S. Hoveland, pp. 109–27. Madison, Wisconsin: American Society of Agronomy.

Jagusch, K. T. (1973). Livestock production from pasture. In *Pastures and Pasture Plants*, ed. R. H. M. Langer, pp. 229–42. Auckland: A. H. & A. W. Read.

Jamieson, W. S. & Hodgson, J. (1979). The effect of daily herbage allowance and sward characteristics upon the ingestive behaviour and herbage intake of calves under strip-grazing management. *Grass and Forage Science*, **34**, 261–71.

Janzen, D. H. (1969). Seed-eaters versus seed size, number toxicity and dispersal. *Evolution*, **23**, 1–27.

Janzen, D. H. (1970). Herbivores and the number of tree species in tropical forests. *American Naturalist*, **104**, 501–28.

Jarrige, R., Demarquilly, C. & Dulphy, J. P. (1982). Forage conservation. In *Nutritional Limits to Animal Production from Pastures*, ed. J. B. Hacker, pp. 363–87. Slough, UK: Commonwealth Agricultural Bureaux.

Jewiss, O. R. (1972). Tillering in grasses – its significance and control. *Journal of the British Grassland Society*, **27**, 65–82.

Johnson, C. B. (ed.) (1981). *Physiological Processes Limiting Plant Productivity*. 395 pp. London: Butterworth.

Jollans, J. L. (ed.) (1981). *Grassland in the British Economy*. CAS Paper 10. 589pp. Reading, UK: Centre for Agricultural Strategy.

Jones, C. A. (1983). Effect of soil texture on critical bulk densities for root growth. *Soil Science Society of America Journal*, **47**, 1208–11.

Jones, M. B. & Woodmansee, R. G. (1979). Biogeochemical cycling in annual grassland ecosystems. *Botanica Review*, **45**, 111–44.

Jones, R. J. (1985). Leucaena toxicity and the ruminal metabolism of mimosine. In *Plant Toxicology*, ed. A. A. Seawright, M. P. Hegarty, L. F. James & R. F. Keeler, pp. 111–19. Brisbane: Queensland Government Printer.

Jones, R. J. & Sandland, R. L. (1974). The relation between animal gain and stocking rate. *Journal of Agricultural Science*, **83**, 335–42.

Jones, R. M. (1980). Survival of seedlings and primary taproots of white clover (*Trifolium repens*) in subtropical pastures in south-east Queensland. *Tropical Grasslands*, **14**, 19–22.

Jones, R. M. (1986). Persistencia de las especies forrajeras bajo pastoreo. In *De Evaluacion de Pasturas con Animales*, ed. C. Lascano & G. Pizarro, pp. 167–200. Cali, Colombia: Centro International Agricultura Tropical.

Jones, R. M. & Ratcliff, D. (1983) Patchy grazing and its relation to deposition of cattle dung pats in pastures in coastal sub-tropical Queensland. *Journal of the Australian Institute of Agricultural Science*, **49**, 109–11.

Jorgensen, N. A. & Lu, C. D. (1985). Wet fractionation of legume forages. In *Forage Legumes for Energy – efficient Animal Production*, ed. R. F. Barnes, P. R. Ball, R. W. Brougham, G. C. Marten & D. J. Minson, pp. 225–32. Washington: US Department of Agriculture Agricultural Research Service.

Judson, G. J., Caple, I. W., Langlands, J. P. & Pete, D. W. (1987). Mineral nutrition of grazing ruminants in southern Australia. In *Temperate Pastures: their Production, Use and Management*, ed. J. L. Wheeler, C. J. Pearson & G. E. Robards, pp. 377–85. Melbourne: Australian Wool Corporation and Commonwealth Scientific and Industrial Research Organization.

Kaiser, A. G. & Curll, M. L. (1987). Improving the efficiency of forage conservation from pastures. In *Temperate Pastures: their Production, Use and Management*, ed. J. L. Wheeler, C. J. Pearson & G. E. Robards, pp. 397–411. Melbourne: Australian Wool Corporation and Commonwealth Scientific and Industrial Research Organization.

Karlovsky, J. (1983). Phosphorus utilization in grassland ecosystems. *Proceedings of the XIVth International Grasslands Congress*, pp. 279–81. Boulder, Colorado: Westview Press.

Kay, M. (1976). Meeting the energy and protein requirements of the growing animal. In *Principles of Cattle Production*, ed. H. Swan & H. Broster, pp. 255–70. London: Butterworth.

Kemp, A. & Geurink, J. H. (1978). Grassland farming and minerals in cattle. *Netherlands Journal of Agricultural Science*, **26**, 161–9.

Kemp, D. R. (1984). Temperate pastures. In *Control of Crop Productivity*, ed. C. J. Pearson. pp. 159–84. Sydney: Academic Press.

Kennedy, A. P. & Till, A. R. (1981). The distribution in soil and plant of ^{32}S from sheep excreta. *Australian Journal of Agricultural Research*, **32**, 339–51.

Kenney, P. A. & Black, J. L. (1984). Factors affecting diet selection by sheep. I. Potential intake rate and acceptability of feed. *Australian Journal of Agricultural Research*, **35**, 551–63.

Ketellapper, H. J. (1969). Growth and development in *Phalaris* III. Effect of temperature on critical daylength in geographic strains of *P. tuberosa* L. *Plant and Cell Physiology*, **10**, 461–3.

King, J., Grant, S. A., Torvell, L. & Sim, E. M. (1984). Growth rate, senescence and photosynthesis of ryegrass swards cut to maintain a range of values for leaf area index. *Grass and Forage Science*, **39**, 371–80.

Kingsbury, J. M. (1983). The evolutionary and ecological significance of plant toxins. In *Handbook of Natural Toxins. Plant and Fungal Toxins*, ed. R. F. Keeler & A. T. Tu, pp. 675–706. New York: Marcel Dekker.

Kirkby, E. A. & Knight, A. H. (1977). Influence of the level of nitrate nutrition on ion uptake and assimilation, organic acid accumulation, and cation–anion balance in whole tomato plants. *Plant Physiology*, **60**, 349–53.

Klebs, G. (1885). Beitrage zur Morphologie und Biologie der Keimung. *Untersuchungen Botanische Institut Tubingen*, **1**, 536–635. Cited by Vogel (1980).

Kleiber, M. (1961). *The Fire of Life*: New York: John Wiley & Sons.

Klinner, W. E. & Shepperson, G. (1975). The state of haymaking technology. *Journal of the British Grassland Society*, **30**, 259–66.

Kothmann, M. M. & Smith, G. M. (1983). Evaluating management alternatives with a beef production system model. *Journal of Range Management*, **36**, 733–40.

Krylova, N. P. (1979). Seed propagation of legumes in natural meadows of the USSR – Review. *Agro-Ecosystems*, **5**, 1–22.

Küntzel, U. (1982). Biogas production from green crops. In *Efficient Grassland farming*, ed. A. J. Corrall, p. 321. Hurley, UK: British Grassland Society.

Lacey, J. R. & Poollen, van, H. W. (1981). Comparison of herbage production on moderately grazed and ungrazed western ranges. *Journal of Range Management*, **34**, 210–12.

Lancashire, J. A. (ed.) (1980). *Herbage Seed Production*. Grassland Research & Practise Series No. 1. 122 pp. Palmerston North: New Zealand Grassland Association.

Landsberg, J. J. (1977). Effects of weather on plant development. In *Environmental Effects on Crop Physiology*, ed. J. J. Landsberg & C. V. Cutting, pp. 289–308. London: Academic Press.

Langer, R. H. M. (1956). Growth and nutrition of timothy (*Phleum pratense*) I. the life history of individual tillers. *Annals of Applied Biology*, **44**, 166–87.

Langer, R. H. M. (1980). Growth of the grass plant in relation to seed production. In *Herbage Seed Production*, ed. J. A. Lancashire, pp. 6–11. Palmerston North: New Zealand Grassland Association.

Latch, G. C. M., Hunt, W. F. & Musgrave, D. R. (1985). Endophytic fungi affect growth of perennial ryegrass. *New Zealand Journal of Agricultural Research*, **28**, 165–8.

Leach, G. (1976). *Energy and Food Production*. 137 pp. Guildford, UK: IPC Science and Technology Press.

Leach, G. J. (1978). The ecology of lucerne pastures. In *Plant Relations in Pastures*, ed. J. R. Wilson, pp. 290–308. Melbourne: Commonwealth Scientific and Industrial Research Organization.

Leafe, E. L. & Parsons, A. J. (1983). Physiology of growth of a grazed sward. *Proceedings of the XIVth International Grasslands Congress*, pp. 403–6. Boulder, Colorado: Westview Press.

Lee, H. J., Kuchel, R. E., Good, B. F. & Trowbridge, R. F. (1957). The aetiology of phalaris staggers in sheep. III. The preventative effect of various oral dose rates of cobalt. *Australian Journal of Agricultural Research*, **8**, 494–501.

Lehr, J. J. & van Weismael, J. G. (1961). Volatilization of ammonia on calcareous soils. *Landbouwsckool Tijdschreift*, **73**, 1156–68.

Leith, H. (1960). Patterns of change within grassland communities. In *The Biology of Weeds*, ed. J. L. Harper, pp. 27–39. Oxford: Blackwell.

Leith, H. (1973). Primary production: terrestrial ecosystems. *Human Ecology*, **1**, 303–32.

Leng, R. A. (1987). Determining the nutritive value of herbages. In *Forages in South East Asian and South Pacific Agriculture*, ed. G. J. Blair, D. A. Ivory & T. R. Evans, pp. 111–24 Canberra: Australian Centre for International Agricultural Research.

Lenné, J. M., Turner, J. W. & Cameron, D. F. (1980). Resistance to diseases and pests of tropical pasture plants. *Tropical Grasslands*, **14**, 146–52.

Leslie, J. K. (1965). Factors responsible for failure in the establishment of summer grasses on the black earths of the Darling Downs, Queensland. *Queensland Journal of Agricultural and Animal Sciences*, **22**, 17–38.

Liebholz, J. & Kellaway, R. C. (1984). The utilization of low quality roughages. 1. The role of nitrogen and energy supplements. *Australian Meat Research Committee Review*, **48**, 1–21.

Linacre, E. (1968). Estimating the net radiation flux. *Agricultural Meteorology*, **5**, 49–63.

List, R. J. (ed.) (1958). *Smithsonian Meteorological Tables*, 6th edn. 325 pp. Washington: Smithsonian Institution.

Littleton, E. J., Dennett, M. D., Elston, J. & Monteith, J. L. (1979). The growth and development of cowpeas (*Vigna unguiculata*) under tropical field conditions. I. Leaf area. *Journal of Agricultural Science*, **93**, 291–307.

Lloyd-Davies, H. (1987). Limitations to livestock production associated with phyto-oestrogens and bloat. In *Temperate Pastures: their Production, Use and Management*, ed. J. L. Wheeler, C. J. Pearson & G. E. Robards, pp. 446–56. Melbourne: Australian Wool Corporation and Commonwealth Scientific and Industrial Research Organization.

Lockwood, J. G. (1974). *World Climatology. An Environmental Approach*. 330 pp. London: Edward Arnold.

Lodha, B. C. (1974). Decomposition of digested litter. In *Biology of Plant Litter Decomposition*, Vol. 1, ed. C. H. Dickinson & G. J. F. Pugh, pp. 213–41. London: Academic Press.

Lovato, A. (1981). Germination of seeds. *Advances in Research and Technology of Seeds*, **6**, 86–120.

Low, S. G. (1984). Ensilage and storage of by-products. In *Silage in the 80's*, ed. T. J. Kempton, A. G. Kaiser & T. E. Trigg. 283 pp. Armidale, Australia: T. J. Kempton.

Lowrance, R., Stinner, B. R. & House, G. J. (eds.) (1984). *Agricultural Ecosystems: Unifying Concepts*. 233 pp. New York: John Wiley and Sons.

Ludlow, M. M. (1980). Stress physiology of tropical pasture plants. *Tropical Grasslands*, **14**, 136–45.

Lym, R. G. & Massersmith, C. G. (1985). Leafy spurge control and improved forage production with herbicides. *Journal of Range Management*, **38**, 386–91.

Lyttleton, J. W. (1973). Proteins and nucleic acids. In *Chemistry and Biochemistry of Herbage*, Vol. I, ed. G. W. Butler & R. W. Bailey, pp. 63–103. London: Academic Press.

McBarron, E. J. (1976). *Medical and Veterinary Aspects of Plant Poisons in New South Wales*. 243 pp. Sydney: NSW Department of Agriculture.

McCalla, T. M. & Haskins, F. A. (1964). Phytotoxic substances from soil, microorganisms and crop residues. *Bacteriological Review*, **28**, 181–6.

McCown, R. L., Jones, R. K. & Peake, D. C. I. (1985). Evaluation of a no-till, tropical legume ley-farming strategy. In *Agro-Research for the Semi-Arid Tropics: North-West Australia*, ed. R. C. Muchow, pp. 450–72. St Lucia, Australia: University of Queensland Press.

McCown, R. L., Wall, B. H. & Harrison, P. G. (1981). The influence of weather on the quality of tropical legume pasture during the dry season in northern Australia: I Trends in sward structure and moulding of standing hay at three locations. *Australian Journal of Agricultural Research*, **32**, 575–87.

McDonald, P. (1981). *The Biochemistry of Silage*. 226 pp. Chichester, UK: John Wiley & Sons.

McDonald, P., Edwards, R. A. & Greenhalgh, J. F. D. (1975). *Animal Nutrition*, 2nd edn. 479 pp. London: Longman.

McDowell, R. E. (1985). Meeting constraints to livestock production systems in Asia. *Proceedings of a Regional Workshop on Livestock Production Management*, pp. 37–66. Manila: Asian Development Bank.

McDowell, R. E. & Hildebrand, P. E. (1980). *Integrated Crop and Animal Production: Making the Most of Resources Available to Small Farms in Developing Countries*. The Rockefeller Foundation, USA.

Macfadyen, A. (1971). *Animal Ecology. Aims and Methods*, 2nd edn. 344 pp. London: Pitman.

McIvor, J. C. & Bray, R. A. (eds.) (1983). *Genetic Resources of Forage Plants*. Melbourne: Commonwealth Scientific and Industrial Research Organization.

McKenzie, J. (1981). Changing consumer demand. In *Grassland in the British Economy*, ed. J. L. Jollans, pp. 107–17. Reading, UK: Centre for Agricultural Strategy.

McKeon, G. M. (1984). Field changes in germination requirements: effect of natural rainfall on potential germination speed and light requirement of *Stylosanthes humilis, Stylosanthes hamata* and *Digitaria ciliaris*. *Australian Journal of Agricultural Research*, **35**, 807–20.

McKeon, G. M. & Mott, J. J. (1984). Seed biology of *Stylosanthes*. In *The Biology and Agronomy of Stylosanthes*, ed. H. M. Stace & L. A. Edye, pp. 311–32. Sydney: Academic Press.

McKeon, G. M. & Rickert, K. G. (1986). Tropical pastures in the farming system: case studies of modelling integration through simulation. In *Proceedings of the 3rd Australian Conference on Tropical Pastures*, ed. G. J. Murtagh & R. M. Jones, pp. 92–100. Brisbane: Tropical Grassland Society of Australia.

McLeod, M. N. (1974). Plant tannins – their role in forage quality. *Nutrition Abstracts and Reviews*, **44**, 803–15.

McManus, W. R., Robinson, V. N. E. & Grout, L. T. (1977). The physical distribution of mineral material on forage plant cell walls. *Australian Journal of Agricultural Research*, **28**, 651–62.

McMeekan, C. P. & Walshe, M. J. (1963). Inter-relationships of grazing methods and stocking rate in the efficiency of pasture utilization by dairy cattle. *Journal of Agricultural Science*, **61**, 147–63.

McWilliam, J. R. (1978). Response of pasture plants to temperature. In *Plant Relations in Pastures*, ed. J. R. Wilson, pp. 17–34. Melbourne: Commonwealth Scientific and Industrial Research Organization.

McWilliam, J. R., Clements, R. J. & Dowling, P. M. (1970). Some factors influencing the germination and early seedling development of pasture plants. *Australian Journal of Agricultural Research*, **21**, 19–32.

McWilliam, J. R. & Dowling, P. M. (1970). Factors influencing the germination and establishment of pasture seed on the soil surface. *Proceedings of the 11th International Grassland Congress*, pp. 578–82. St Lucia, Queensland: University of Queensland Press.

Madden, J. P. (1985). *Beyond Conventional Economics*. Paper Series, College of Agriculture, The Pennsylvania State University. 52 pp.

Maelzer, D. A. (1962). The emergence as a pest of *Aphodius tasmaniae* Hope in pastures in the lower South East of South Australia. *Australian Journal of Zoology*, **10**, 95–112.

MAF (Ministry of Agriculture and Fisheries) (1976). *Nutrient Allowances and Composition of Feedstuffs for Ruminants*. Wellington: LGR21 Nutrition Chemists Publication Committee, New Zealand Government Printer.

MAFF (Ministry of Agriculture, Fisheries and Food) (1975). *Energy Allowances and Feeding Systems for Ruminants*. 79 pp. Technical Bulletin 33. London: HMSO (Her Majesty's Stationery Office).

Mahon, J. D. (1983). Energy relationships. In *Nitrogen Fixation*, Vol. 3, *Legumes*, ed. W. J. Broughton, pp. 299–326. Oxford: Clarendon Press.

Malberg, C. & Smith, H. (1982). Relationship between plant weight and density in mixed populations of *Medicago sativa* and *Trifolium pratense*. *Oikos*, **38**, 365–8.

Margaleff, R. (1969). Diversity and stability: a practical proposal and a model of interdependence. In *Diversity and Stability of Ecological Systems*, pp. 25–37. Brookhaven Symposia in Biology 22. Springfield, Virginia: National Technical Information Service.

Marsh, J. S. (1981). The politics of grass. In *Grassland in the British Economy*, ed. J. L. Jollans, pp. 64–75. CAS Paper 10. Reading, UK: Centre for Agricultural Strategy.

Marsh, R. & Campling, R. C. (1970). Fouling of pastures by dung. *Herbage Abstracts*, **40**, 123–30.

Marshall, C. & Sagar, G. R. (1965). The influence of defoliation on the distribution of assimilates in *Lolium multiflorum* Lam. *Annals of Botany*, **29**, 365–70.

Marshall, D. R. & Broue, P. (1973). Outcrossing rates in Australian populations of subterranean clover. *Australian Journal of Agricultural Research*, **24**, 863–7.

Marten, G. C. & Andersen, R. N. (1975). Forage nutritive value and palatability of 12 common annual weeds. *Crop Science*, **15**, 821–57.

Martin, P. B. (1983). Insect habitat management in pasture systems. *Environmental Management*, **7**, 59–64.

May, L. H. (1960). The utilization of carbohydrate reserves in pasture plants after defoliation. *Herbage Abstracts*, **30**, 239–45.

May, P. F., Till, A. R. & Cumming, M. J. (1972). Systems analysis of ^{35}sulphur kinetics in pastures grazed by sheep. *Journal of Applied Ecology*, **9**, 25–49.

Mears, P. T. (1970). Kikuyu (*Pennisetum clandestinum*) as a pasture grass. A review. *Tropical Grasslands*, **4**, 139–52.

Mengel, K. & Kirkby, E. A. (1979). *Principles of Plant Nutrition*, 2nd edn. Berne: International Potash Institute.

Middleton, K. R. & Smith, C. S. (1978). The concept of a climax in relation to the fertilizer input of a pastoral ecosystem. *Plant and Soil*, **50**, 595–614.

Milford, R. & Minson, D. J. (1966). Intake of tropical pasture species. *Proceedings of the 9th International Grassland Congress*, pp. 815–22.

Miller, A. (1985). Cognitive styles and environmental problem-solving. *International Journal of Environmental Studies*, **26**, 21–31.

Milligan, K. (1984). Comments on the grazing management session. *Proceedings of the New Zealand Grassland Association*, **45**, 207–8.

Milligan, K. E. & McConnell, G. R. (1976). *Feed Budgeting*. Wellington, New Zealand: Ministry of Agriculture and Fisheries.

Milthorpe, F. L. & Moorby, J. (1979). *An Introduction to Crop Physiology*, 2nd edn. 244 pp. Cambridge University Press.

Minchin, F. R. & Pate, J. S. (1973). The carbon balance of a legume and the functional economy of its root nodules. *Journal of Experimental Botany*, **24**, 259–71.

Minson, D. J. (1982). Effects of chemical and physical

composition of herbage eaten upon intake. In *Nutritional Limits to Animal Production from Pastures*, ed. J. B. Hacker, pp. 167–82. Farnham Royal, UK: Commonwealth Agricultural Bureaux.

Minson, D. J., Harris, C. E., Raymond, W. F. & Milford, R. (1964). The digestibility and voluntary intake of S22 meadow fescue and Germinal cocksfoot. *Journal of the British Grassland Society*, **19**, 298–305.

Mitchell, A. & King, P. N. (1980). Land use and management of water supply catchments. *Proceedings of the First Australian Agronomy Conference*, pp. 40–53. Brisbane: The Australian Society of Agronomy.

Moir, K. W., Wilson, J. R. & Blight, G. W. (1977). The *in vitro* digested cell wall and fermentation characteristics of grasses as affected by temperature and humidity during their growth. *Journal of Agricultural Science*, **88**, 217–22.

Moncur, M. W. (1981). *Floral Initiation in Field Crops. An Atlas of Scanning Electron Micrographs*. 135 pp. Melbourne: Commonwealth Scientific and Industrial Research Organization.

Monteith, J. L. (1972). Solar radiation and productivity in tropical ecosystems. *Journal of Applied Ecology*, **9**, 747–66.

Monteith, J. L. (1981). Does light limit crop production? In *Physiological Processes Limiting Plant Productivity*, ed. C. B. Johnson, pp. 23–38. London: Butterworth.

Monteith, J. L. & Elston, J. (1983). Performance and productivity of foliage in the field. In *The Growth and Functioning of Leaves*, ed. J. E. Dale & F. L. Milthorpe, pp. 499–518. Cambridge University Press.

Mooney, H. A. (1972). The carbon balance of plants. *Annual Review of Ecology and Systematics*, **3**, 315–46.

Moore, C. W. E. (1966). Distribution of grasslands. In *Grasses and Grasslands*, ed. C. Barnard, pp. 182–205. New York: Macmillan.

Morgan, R. P. C. (1985). Priorities for technical research in soil conservation. *Food and Fertilizer Technology Centre Extension Bulletin*, **218**, 1–9. Taipei City, Taiwan: FFTC.

Morley, F. H. W. (1981). Management of grazing systems. In *Grazing Animals*, ed. F. H. W. Morley, pp. 379–400. Amsterdam: Elsevier.

Morley, F. H. W. (1987). The economics of pasture development and pasture research. In *Temperate Pastures: their Production, Use and Management*, ed. J. L. Wheeler, C. J. Pearson & G. E. Robards, pp. 571–9. Melbourne: Australian Wool Corporation and Commonwealth Scientific and Industrial Research Organization.

Morley, F. H. W. & Donald, A. D. (1980). Farm management and systems of helminth control. *Veterinary Parasitology*, **6**, 105–34.

Morley, F. H. W. & White, D. H. (1985). Modelling biological systems. In *Agricultural Systems Research for Developing Countries*, ed. J. V. Remenyi, pp. 60–9. Canberra: Australian Centre for International Agricultural Research.

Mortimer, A. M. (1974). Studies of germination and establishment of selected species with special reference to the fate of seeds. PhD thesis, University College of North Wales. (Cited in Grime, 1979.)

Mortimer, P. H. & Menna, M. B., di (1985). Interactions of *Lolium* endophyte on pasture production and perennial ryegrass staggers disease. In *Trichothecenes and Other Mycotoxins*, ed. J. Lacey, pp. 00–00. London: John Wiley & Sons.

Mott, G. O. (1960). Grazing pressure and the measurement of pasture production. *Proceedings of the 8th International Grassland Congress*, pp. 606–11. Oxford: Alden Press.

Mott, J. J. (1978). Dormancy and germination in five native grass species from savanna woodland communities of the Northern Territory. *Australian Journal of Botany*, **26**, 621–31.

Mott, J. J. (1979). High temperature contact treatment of hard seed in *Stylosanthes*. *Australian Journal of Agricultural Research*, **30**, 847–54.

Mott, J. J., Otthill, J. C. & Weston, E. J. (1981). Animal production from the native woodlands and grasslands of northern Australia. *Journal of the Australian Institute of Agricultural Science*, **47**, 132–46.

Muchow, R. C. (ed.) (1985). *Agro-Research for the Semi-arid Tropics: North-West Australia*. 608 pp. St Lucia, Australia: University of Queensland Press.

Muir, A. M. (1985). *Livestock and Pasture Budgets for Central-Western New South Wales*. 89 pp. Department of Agriculture, NSW: Agdex 815.

Muldoon, D. K. (1979). Simulation of hybrid pennisetum production in Australia. *Agricultural Systems*, **4**, 39–47.

Muldoon, D. K. & Pearson, C. J. (1979). Morphology and physiology of regrowth of the tropical tall grass hybrid pennisetum. *Annals of Botany*, **43**, 719–28.

Muldoon, D. K., Wheeler, J. L. & Pearson, C. J. (1984). Growth, mineral composition and digestibility of maize, sorghum and barnyard millets as influenced by temperature. *Australian Journal of Agricultural Research*, **35**, 367–78.

Mulholland, J. G., Coombe, J. B. & Pearce, G. R. (1984). Conservation and utilization of crop residues. In *Silage in the 80's*, ed. T. J. Kempton, A. G. Kaiser & T. E. Trigg, pp. 301–16. Armidale, Australia: T. J. Kempton.

Munns, D. N. (1978). Soil acidity and nodulation. In *Mineral Nutrition of Legumes in Tropical and Subtropical Soils*, ed. C. S. Andrews & E. J. Kamprath, pp. 247–64. Melbourne: Commonwealth Scientific and Industrial Research Organization.

Murray, D. R. (ed.) (1984). *Seed Physiology*, Vol. 2. 259 pp. Sydney: Academic Press.

Murtagh, G. J. (1980). Integration of feed sources in property management: intensive systems. *Tropical Grasslands*, **14**, 232–8.

Myers, R. J. K. & Henzell, E. F. (1985). Productivity and economics of legume-based versus nitrogen-fertilized grass-based forage systems in Australia. In *Forage Legumes for Energy-Efficient Animal Production*, ed. R. F. Barnes, P. R. Ball, R. W. Brougham, G. C. Martin & D. J. Minson, pp. 40–6. Springfield, Virginia: US Department of Agriculture Agricultural Research Service.

Naylor, R. E. L., Marshall, A. H. & Matthews, S. (1983). Seed establishment in directly drilled sowings. *Herbage Abstracts*, **53**, 73–91.

Nelson, C. J., Asay, K. H. & Horst, G. L. (1975). Relationship of leaf photosynthesis to forage yield of tall fescue. *Crop Science*, **15**, 476–8.

Nestel, B. (ed.) (1984). *Development of Animal Production Systems*. 435 pp. Amsterdam: Elsevier.

Neto, M. S. & Jones, R. M. (1983). The passage of pasture seed through the ruminant digestive tract. *CSIRO Division of Tropical Crops and Pastures Annual Report*, pp. 100–3. Brisbane: Commonwealth Scientific and Industrial Research Organization.

Newell, S. J. & Tramer, E. J. (1978). Reproductive strategies in herbaceous plant communities during succession. *Ecology*, **59**, 228–34.

Nitis, I. M. (1986). Present state of grassland production and utilization and future perspectives for grassland farming in humid tropical Asia. *Proceedings of the XVth International Grassland Congress*, pp. 39–44. Nishi-nasuno, Japan: The Japanese Society of Grassland Science.

Nolan, T. & Connolly, J. (1977). Mixed stocking by sheep and steers – a review. *Herbage Abstracts*, **47**, 367–74.

Norman, D. & Collinson, M. (1985). Farming systems research in theory and practice. In *Agricultural Systems Research for Developing Countries*, ed. J. V. Remenyi, pp. 16–30. Canberra: Australian Centre for International Agricultural Research.

Norman, M. J. T., Pearson, C. J. & Searle, P. G. E. (1984). *The Ecology of Tropical Food Crops*. 369 pp. Cambridge University Press.

Norton, B. W. (1982). Differences between species in forage quality. In *Nutritional Limits to Animal Production from Pastures*, ed. J. B. Hacker, pp. 89–110. Slough, UK: Commonwealth Agricultural Bureaux.

Noy-Meir, I. (1976). Rotational grazing in a continuously growing pasture: a simple model. *Agricultural Systems*, **1**, 87–112.

Nye, P. H. & Tinker, P. B. (1969). The concept of a root demand coefficient. *Journal of Applied Ecology*, **6**, 293–300.

Oddy, V. H. (1983). *Feed Requirements of Sheep and Cattle During Drought using a Metabolizable Energy System*. 32 pp. Agbulletin 3, Sydney: Department of Agriculture of New South Wales.

Ojima, K. & Isawa, T. (1968). The variation in carbohydrates in various species of grasses and legumes. *Canadian Journal of Botany*, **46**, 1507–11.

Okubo, T. (1982). Energy flow modelling for plant and animal growth in relation to the advantage of a grazing system for a more efficient agriculture. *Proceedings of a Workshop on Green Energy for Regional Development*. 18 pp. Bogor Agricultural University: Bogor, Indonesia.

Okubo, T., Hirakawa, M., Okajima, T. & Kayama, R. (1983). Comparison of radiant energy distribution and conversion efficiency in grazed pastures between temperate grass sward and subtropical grass sward. *Journal of Japanese Grassland Science*, **29**, 73–4.

O'Leary, G. J., Connor, D. J. & White, D. H. (1985). A simulation model of the development, growth and yield of the wheat crop. *Agricultural Systems*, **17**, 1–26.

Ong, C. K. (1983). Response to temperature in a stand of pearl millet (*Pennisetum typhoides*, S. & H.). II. Reproductive development. *Journal of Experimental Botany*, **34**, 337–48.

Ong, C. K., Marshall, C. & Sagar, G. R. (1978). The effects of nutrient supply on flowering and seed production in *Poa annua* L. *Journal of the British Grassland Society*, **33**, 117–21.

Oram, R. N. & Williams, J. D. (1976). Variation in concentration and composition of toxic alkaloids among strains of *Phalaris tuberosa* L. *Nature*, **213**, 946–7.

Orr, R. J. & Newton, J. E. (1984). Grazing cycle length and silage supplementation: effects on the performance of lactating ewes with twin lambs. *Grass and Forage Science*, **39**, 323–9.

Osbourne, D. F. (1980). The feeding value of grass and grass products. In *Grass, its Production and Utilization*, ed. W. Holmes, pp. 70–124. Oxford: Blackwell.

Parsons, A. J., Collett, B. & Lewis, J. (1984). Changes in the structure and physiology of a perennial ryegrass sward when released from a continuous stocking management. Implications for the use of exclusion cages in continuously stocked swards. *Grass and Forage Science*, **39**, 1–9.

Parsons, A. J., Leafe, E. L., Collett, B. & Stiles, W. (1983a). The physiology of grass production under grazing. I. Characteristics of leaf and canopy photosynthesis of continuously-grazed swards. *Journal of Applied Ecology*, **20**, 117–26.

Parsons, A. J., Leafe, E. L., Collett, B., Penning, P. D. & Lewis, J. (1983b). The physiology of grass production under grazing. II. Photosynthesis, crop growth and animal intake of continuously-grazed swards. *Journal of Applied Ecology*, **20**, 127–39.

Parsons, A. J. & Robson, M. J. (1982). Seasonal changes in the physiology of S24 perennial ryegrass (*Lolium perenne* L.) 4. Comparison of the carbon balance of the reproductive crop in spring and the vegetative crop in autumn. *Annals of Botany*, **50**, 167–77.

Pate, J. S. & Atkins, C. A. (1983). Nitrogen uptake, transport and utilization. In *Nitrogen Fixation*, Vol. 3, *Legumes*, ed. W. J. Broughton, pp. 245–98. Oxford: Clarendon Press.

Pearson, C. J. (1979). Daily cycles of photosynthesis, respiration, and translocation. In *Photosynthesis and Plant Development*, ed. R. Marcele, H. Clijsters & M. van Poucke, pp. 125–36. The Hague: Dr. Junk.

Pearson, C. J. (ed.) (1984). *Control of Crop Productivity*. 315 pp. Sydney: Academic Press.

Pearson, C. J. & Hunt, L. A. (1972a). Effects of temperature on primary growth of alfalfa. *Canadian Journal of Plant Science*, **52**, 1007–15.

Pearson, C. J. & Hunt, L. A. (1972b). Studies on the daily course of carbon exchange in alfalfa plants. *Canadian Journal of Botany*, **50**, 1377–84.

Pearson, C. J. & Jacobs, B. C. (1985). Root distribution in space and time in *Trifolium subterraneum*. *Australian Journal of Agricultural Research*, **36**, 601–14.

Pearson, C. J., Kemp, J., Kirby, A. C., Launders, T. E. & Mikled, C. (1985). Responsiveness to seasonal temperature and nitrogen among genotypes of kikuyu, paspalum and bermuda grass pastures in coastal New South Wales. *Australian Journal of Experimental Agriculture*, **25**, 109–16.

Peel, S. & Matkin, E. A. (1984). Herbage yield and animal production from grassland on three commercial dairy farms in south-east England. *Grass and Forage Science*, **39**, 177–85.

Penman, H. L. (1963). *Vegetation and Hydrology*. Technical Communication 53. Farnham Royal: Commonwealth Agricultural Bureaux.

Perkins, J., Petheram, J., Rachman, R. & Semali, A. (1986). Prospects for the introduction and management of forages in livestock production systems of South-East Asia and the South Pacific. In *Forages in South East Asian and South Pacific Agriculture*, ed. G. J. Blair, D. A. Ivory & T. R. Evans, pp. 15–23. Canberra: Australian Centre for International Agricultural Research.

Perry, D. A. (1976). Seed vigour and seedling establishment. *Advances in Research and Technology of Seeds*, **2**, 62–85.

Peters, E. J. (1976). Aerial application of herbicides, seed and fertilizer improves forage on Ozark hill lands. In *Hill Lands*, ed. J. Luchok, S. D. Cawthon & M. J. Breslin, pp. 167–70. Morgantown, West Virginia, USA: West Virginia University Press.

Peterschmidt, N. A., Delaney, R. H. & Greene, M. C. (1979). Effects of overirrigation on growth and quality of alfalfa. *Agronomy Journal*, **71**, 752–4.

Phillips, D. A. (1980). Efficiency of symbiotic nitrogen fixation in legumes. *Annual Review of Plant Physiology*, **31**, 29–49.

Pieper, R. D. (1983). Consumption rates of desert grassland herbivores. *Proceedings of the XIVth International Grassland Congress*, pp. 465–7. Boulder, Colorado: Westview Press.

Pimentel, D. (ed.) (1980). *Handbook of Energy Utilization in Agriculture*. Boca Raton, Florida: CRC Press.

Plowright, R. C. & Hartling, L. K. (1981). Red clover pollination by bumble bees: a study of the dynamics of a plant-pollinator relationship. *Journal of Applied Ecology*, **18**, 639–47.

Plucknett, D. L. (1979). *Managing Pastures and Cattle Under Coconuts*. 364 pp. Boulder, Colorado: Westview Press.

Pook, E. W. & Costin, A. B. (1971). Changes in pattern and density of perennial grasses in an intensively grazed sown pasture influenced by drought in southern New South Wales. *Australian Journal of Experimental Agriculture and Animal Husbandry*, **10**, 286–92.

Pott, A. & Humphreys, L. R. (1983). Persistence and growth of *Lotononis bainesii* – *Digitaria decumbens* pasture. 1. Sheep stocking rate. *Journal of Agricultural Science*, **101**, 1–7.

Pott, A., Humphreys, L. R. & Hales, J. W. (1983). Persistence and growth of *Lotononis bainesii* – *Digitaria decumbens* pastures 2. Sheep treading. *Journal of Agricultural Science*, **101**, 9–15.

Preston, F. W. (1969). Diversity and stability in the biological world. In *Diversity and Stability of Ecological Systems*, pp. 1–12. Brookhaven Symposia in Biology 22. Springfield, Virginia: National Technical Information Service.

Preston, R. L. (1972). Protein requirements for growing and lactating ruminants. *Proceedings of the University of Nottingham 6th Nutritional Conference of Feed Manufacturers*, p. 22. (Cited by Kay, 1976.)

Preston, T. R. (1984). New approaches to animal nutrition in the tropics. In *Development of Animal Production Systems*, ed. B. Nestel, pp. 379–96. Amsterdam: Elsevier.

Probert, M. E. (1984). The mineral nutrition of *Stylosanthes*. In *The Biology and Agronomy of Stylosanthes*, ed. H. M. Stace & L. A. Edye, pp. 203–26. Sydney: Academic Press.

Purvis, G. & Curry, J. P. (1981). The influence of sward management on foliage arthropod communities in a ley grassland. *Journal of Applied Ecology*, **18**, 711–25.

Rabinowitz, D. & Rapp, J. K. (1980). Seed rain in a North American tall grass prairie. *Journal of Applied Ecology*, **17**, 793–802.

Ralphs, M. H. & Busby, F. E. (1979). Prescribed burning: vegetative change, forage production, cost and returns on six demonstration burns in Utah. *Journal of Range Management*, **32**, 267–70.

Ranjhan, S. K. & Chadhoker, P. A. (1984). Effective utilization of agro-industrial by-products for animal feeding in Sri Lanka. *World Animal Review*, **50**, 45–51.

Raymond, F., Shepperson, G. & Waltham, R. (1978). *Forage Conservation and Feeding*. 208 pp. Ipswich, UK: Farming Press Ltd.

Read, J. W. (1976). Some factors affecting production of two species and two interspecific hybrids of *Phalaris* under irrigation. Master of Rural Science, University of New England, Armidale, Australia.

Reed, K. F. M. & Cocks, P. S. (1982). Some limitations of pasture species in southern Australia. *Proceedings of the 2nd Agronomy Conference , Wagga Wagga, Australia*, pp. 142–60. Sydney: Australian Society of Agronomy.

Reeves, T. G. (1987). Pastures in cropping systems. In *Temperate Pastures: their Production, Use and Management*, ed. J. L. Wheeler, C. J. Pearson & G. E. Robards, pp. 501–15. Melbourne: Australian Wool Corporation and Commonwealth Scientific and Industrial Research Organization.

Reid, R. L. & Jung, G. A. (1982). Problems of animal production from temperate pastures. In *Nutritional Limits to Animal Production from Pastures*, ed. J. B. Hacker, pp. 21–43. Slough, UK: Commonwealth Agricultural Bureaux.

Rhodes, I. (1971). The relationship between productivity and components of canopy structure in ryegrass (*Lolium* spp.) II. *Journal of Agricultural Science*, **77**, 283–92.

Rhodes, I. (1972). Yield, leaf area index and photosynthetic rate in some perennial ryegrass (*Lolium perenne* L.) selections. *Journal of Agricultural Science*, **28**, 509–11.

Rika, I. K. (1986). The role and management of forages in plantation crops. In *Forages in South East Asian and South Pacific Agriculture*, ed. G. J. Blair, D. A. Ivory & T. R. Evans, pp. 158–61. Canberra: Australian Centre for International Agricultural Research.

Robards, G. E., Leigh, J. H. & Mulham, W. E. (1967). Selection of diet by sheep grazing semi-arid pastures on the Riverina Plain 4. A grassland (*Danthonia caespitosa*) community. *Australian Journal of Experimental Agriculture and Animal Husbandry*, **7**, 426–33.

Roberts, E. H. (1972). Dormancy: a factor affecting seed survival in the soil. In *Viability of Seeds*, ed. E. H. Roberts, pp. 321–59. London: Chapman and Hall.

Roberts, H. A. (1981). Seed banks in soils. *Advances in Applied Biology*, **6**, 1–56.

Roberts, R. J. (1979). Insect damage assessment and survey techniques. *Proceedings of the 2nd Australasian Conference on Grassland Invertebrate Ecology*, pp. 104–7. Palmerston North, New Zealand.

Robson, A. D. (1983). Mineral nutrition. In *Nitrogen Fixation*, Vol. 3, *Legumes*, ed. W. J. Broughton, pp. 36–55. Oxford: Clarendon Press.

Robson, M. J. (1980). A physiologist's approach to raising the potential yield of the grass crop through breeding. In *Opportunities for Increasing Crop Yields*, ed. R. G. Hurd, P. V. Briscoe & C. Dennis, pp. 33–49. Boston: Pitman Press.

Robson, M. J. (1982). The growth and carbon economy of selected lines of *Lolium perenne* cv. S23 with differing rates of dark respiration 2. *Annals of Botany*, **49**, 331–9.

Rodell, C. F. (1978). Simulation of grasshopper populations in a grassland ecosystem. In *Grassland Simulation Model*, ed. G. S. Innis, pp. 127–54. Ecological Studies 26. New York: Springer Verlag.

Rolston, M. P. (1978). Water impermeable seed dormancy. *Botanical Review*, **44**, 365–89.

Ross, G. D. (1984). The microbiology of silage. In *Silage in the 80's*, ed. T. J. Kempton, A. G. Kaiser, & T. E. Trigg, pp. 94–105. Armidale, Australia: T. J. Kempton.

Ross, P. J., Henzell, E. F. & Ross, D. R. (1972). Effects of nitrogen and light in grass–legume pastures – a systems analysis approach. *Journal of Applied Ecology*, **9**, 535–56.

Rossiter, R. C. (1961). The influence of defoliation on the components of seed yield in swards of subterranean clover (*Trifolium subterraneum* L.). *Australian Journal of Agricultural Research*, **12**, 821–33.

Rossiter, R. C. & Barrow, N. J. (1972). Physiological and ecological studies on the oestrogenic isoflavones in subterranean clover (*T. subterraneum* L.) IX. Effects of sulphur supply. Australian Journal of Agricultural Research, **23**, 411–18.

Rossiter, R. C. & Beck, H. B. (1966). Physiological and ecological studies on the oestrogenic isoflavones in subterranean clover (*T. subterraneum* L.) II. Effects of phosphate supply. *Australian Journal of Agricultural Research*, **17**, 447–56.

Rossiter, R. C., Maller, R. A. & Pakes, A. G. (1985). A model of changes in the composition of binary mixtures of subterranean clover strains. *Australian Journal of Agricultural Research*, **36**, 119–44.

Rosswall, T. & Paustian, K. (1984). Cycling of nitrogen in modern agricultural systems. *Plant and Soil*, **76**, 3–21.

Roux, P. W. & Vorster, M. (1983). Vegetation change in the karoo. *Proceedings of the Grassland Society of Southern Africa*, **18**, 25–9.

Rumbaugh, M. D. (1985). Breeding bloat-safe cultivars of bloat-causing legumes. In *Forage Legumes for Energy-efficient Animal Production*, ed. R. F. Barnes, P. R. Ball, R. W. Brougham, G. C. Marten & D. J. Minson, pp. 238–45. Springfield, Virginia: US Department of Agriculture.

Russell, E. W. (1973). *Soil Conditions and Plant Growth*, 10th edn. London: Longman.

Rykiel, E. J. (1984). Modelling agroecosystems: lessons from ecology. In *Agricultural Ecosystems. Unifying Concepts*, ed. R. Lowrance, B. R. Stinner & G. J. House, pp. 157–78. New York: John Wiley and Sons.

Ryle, G. J. A. (1967). Effects of shading on inflorescence size and development in temperate perennial grasses. *Annals of Applied Biology*, **59**, 297–308.

Ryle, G. J. A., Powell, C. E. & Gordon, A. J. (1985). Defoliation in white clover: nodule metabolism, nodule growth and maintenance, and nitrogenase functioning during growth and regrowth. *Annals of Botany*, **57**, 263–72.

Salisbury, P. A. & Halloran, G. M. (1983). Flowering sequence and the ordering of impermeability breakdown in seed of subterranean clover (*Trifolium subterraneum*). *Annals of Botany*, **52**, 679–88.

Sambo, E. Y. (1983). Comparative growth of the Australian temperate pasture grasses: *Phalaris tuberosa* L., *Dactylis glomerata* L. and *Festuca arundinacea* Schreb. *New Phytologist*, **93**, 89–104.

Sanchez, P. A. & Salinas, J. G. (1981). Low input management technology for managing Oxisols and Ultisols in tropical America. *Advances in Agronomy*, **34**, 279–406.

Sandow, J. D. (1983). *Insect Pests in Lucerne*. Farmnote 93/83. 4 pp. Perth: Department of Agriculture, Western Australia.

Sauvant, D., Chapoutot, P. & Lapierre, O. (1983). A general model of diet least-cost formulation for lactating and growing cattle. In *Feed Information and Animal Production*, ed. G. E. Robards & R. G. Packham, pp. 375–82. Slough, UK: Commonwealth Agricultural Bureaux.

Scowcroft, W. R., Larkin, P. J., Lörz, H. & Fischer, M. (1982). Recent advances – cell culture in plant improvement. *Proceedings of the Second Agronomy Conference*, Wagga Wagga, Australia, pp. 161–7. Sydney: Australian Society of Agronomy.

Sears, P. D. (1960). Grass-clover relationships in New Zealand. *Proceedings of the 8th International Grassland Congress*, pp. 130–2. Oxford: Alden Press.

Shaner, W. W., Philipp, P. F. & Schmehl, W. R. (1982). *Farming Systems Research and Development Guidelines for Developing Countries*. 414 pp. Boulder, Colorado: Westview Press.

Sharma, M. L. (1976). Interaction of water potential and temperature effects on germination of three semi-arid plant species. *Agronomy Journal*, **68**, 390–4.

Shelton, H. M. (1980). Dry season legume forages to follow paddy rice in N. E. Thailand. I. Species evaluation and effectiveness of native Rhizobium for nitrogen fixation. *Experimental Agriculture*, **16**, 57–66.

Shelton, H. M. & Humphreys, L. R. (1971). Effect of variation in density and phosphate supply on seed production of *Stylosanthes humilis*. *Journal of Agricultural Science*, **76**, 325–8.

Sherwood, M. (1983). Rate of nitrogen applied to grassland in animal wastes. *Proceedings of the XIVth International Grasslands Congress*, pp. 347–50. Boulder, Colorado: Westview Press.

Shiel, R. S. & Rimmer, D. L. (1984). Changes in soil structure

and biological activity on soil meadow hay plots at Cockle Park, Northumberland. *Plant and Soil*, **75**, 349–56.

Siebert, B. D. & Hunter, R. A. (1982). Supplementary feeding of grazing animals. In *Nutritional Limits to Animal Production from Pastures*, ed. J. B. Hacker, pp. 409–26. Slough, UK: Commonwealth Agricultural Bureaux.

Silcock, R. G. (1980). Seedling characteristics of tropical pasture species and their implication for ease of establishment. *Tropical Grasslands*, **14**, 174–80.

Silsbury, J. H. (1961). A study of dormancy, survival, and other characteristics in *Lolium perenne* L. at Adelaide, S.A. *Australian Journal of Agricultural Research*, **12**, 1–9.

Simpson, J. R. & Steele, K. W. (1983). Gaseous nitrogen exchanges in grazed pastures. In *Gaseous Loss of Nitrogen from Plant–Soil Systems*, ed. J. R. Freney & J. R. Simpson, pp. 215–36. The Hague: Martinus Nijhoff/Dr W. Junk.

Simpson, R. J., Lambers, H. & Dalling, M. (1983). Nitrogen redistribution during grain growth in wheat (*Triticum aestivum* L.). *Plant Physiology*, **71**, 7–14.

Simpson, R. J. & Culvenor, R. A. (1986). Photosynthesis, carbon partitioning and herbage yield in pastures. In *Temperate Pastures: their Production, Use and Management*, ed. J. L. Wheeler, C. J. Pearson & G. E. Robards. Melbourne: Australian Wool Corporation and Commonwealth Scientific and Industrial Research Organization.

Smetham, M. L. (1973). Grazing management. In *Pastures and Pasture Plants*, ed. R. H. M. Langer, pp. 179–228. Auckland: A. H. & A. W. Reed Ltd.

Smith, F. W. & Verschoyle, M. J. S. (1973). *Foliar Symptoms of Nutrient Disorders in Paspalum dilatatum*. 9 pp. Commonwealth Scientific and Industrial Research Organization Division of Tropical Pastures Technical Paper 14. Canberra: Commonwealth Scientific and Industrial Research Organization.

Smith, G. M. (1983). Simulation models for evaluating the management of ruminants in developing countries. In *Feed Information and Animal Production*, ed. G. E. Robards & R. G. Packham, pp. 463–73. Slough, UK: Commonwealth Agricultural Bureaux.

Smith, R. C. G., Biddiscombe, E. F. & Stern, W. R. (1972). Evaluation of five Mediterranean annual pasture species during early growth. *Australian Journal of Agricultural Research*, **23**, 703–16.

Smith, R. C. G. & Crespo, M. C. (1979). Effect of competition by white clover on the seed production characteristics of subterranean clover. *Australian Journal of Agricultural Research*, **30**, 597–607.

Snaydon, R. W. (1981). 'New' uses for grass. In *Grassland in the British Economy*, ed. J. L. Jollans, pp. 208–26. Reading, UK: Centre for Agricultural Strategy.

Snowball, K. & Robson, A. D. (1983). *Symptoms of Nutrient Deficiencies. Subterranean Clover and Wheat*. 73 pp. Perth: Institute of Agriculture, University of Western Australia.

Soest, P. J., van (1967). Development of a comprehensive system of feed analysis and its application to forages. *Journal of Animal Science*, **26**, 119–28.

Speedy, A. W., Black, W. J. M. & Fitzsimons, J. (1984). The effect of the timing of management actions on the performance of a grassland sheep flock. *Journal of Agricultural Science*, **102**, 275–83.

Spedding, C. R. W. (1971). *Grassland Ecology*. 221 pp. London: Oxford University Press.

Spedding, C. R. W. (1975a). The study of agricultural systems. In *Study of Agricultural Systems*, ed. G. M. Dalton, pp. 3–19. London: Applied Science Publishers Ltd.

Spedding, C. R. W. (1975b) Grazing systems. In *Proceedings of the III World Conference on Animal Production*, ed. R. L. Reid, pp. 145–57. Sydney: Sydney University Press.

Spedding, C. R. W. (1981). The world's grasslands. In *Grassland in the British Economy*, ed. J. L. Jollans, pp. 76–87. Reading, UK: Centre for Agricultural Strategy.

Spedding, C. R. W. (1984*a*). New horizons in animal production. In *World Animal Science 1. Basic Information 2. Development of Animal Production Systems*, ed. B. Nestel, pp. 399–418. Amsterdam: Elsevier.

Spedding, C. R. W. (1984*b*). Agricultural systems and the role of modelling. In *Agricultural Ecosystems. Unifying Concepts*, ed. R. Lowrance, B. R. Stinner & G. J. House, pp. 179–86. New York: Wiley.

Springett, J. A. (1983). Effect of five species of earthworm on some soil properties. *Journal of Applied Ecology*, **20**, 865–72.

Stanhill, G. (ed.) (1984). *Energy and Agriculture*. Berlin: Springer-Verlag.

Stobbs, T. H. (1973*a*). The effect of plant structure on the intake of tropical pastures. I. Variation in the bite size of grazing cattle. *Australian Journal of Agricultural Research*, **24**, 809–19.

Stobbs, T. H. (1973*b*). The effect of plant structure on the intake of tropical pastures. II. Differences in sward structure, nutritive value and bite size of animals grazing *Setaria anceps* and *Chloris gayana* at various stages of growth. *Australian Journal of Agricultural Research*, **24**, 821–9.

Stockdale, C. R. (1985). Influence of some sward characteristics on the consumption of irrigated pastures grazed by lactating dairy cows. *Grass and Forage Science*, **40**, 31–9.

Stockdale, C. R. & King, K. R. (1983). Effect of stocking rate on the grazing behaviour and faecal output of lactating dairy cows. *Grass and Forage Science*, **38**, 215–18.

Stuedemann, J. A., Wilkinson, S. R., Belesky, D. P., Hoveland, C. S., Divine, O. J., Breedlove, D. L., Cordia, H. & Thompson, F. N. (1986). Effect of cultivar level of fungal endophyte (*Acremonium coenophialum*) and nitrogen fertilization of tall fescue (*Festuca arundinacea*) on steer performance. *Proceedings of the XVth International Grassland Congress*, pp. 1171–3. Nishi-nasuno, Japan: The Japanese Society of Grassland Science.

Stynes, B. A. & Bird, A. F. (1983). Development of annual ryegrass toxicity. *Australian Journal of Agricultural Research*, **34**, 653–60.

Suckling, F. E. T. (1976). A 20 year study of pasture development through phosphate and legumes oversowing on North Island hill country of New Zealand. In *Hill Lands*, ed. J. Luchok, J. D. Cawthon & M. J. Breslin, pp. 367–80. Morgantown, West Virginia: West Virginia University Press.

Suckling, F. E. T. & Charlton, J. F. L. (1978). A review of the significance of buried legume seeds with particular reference to New Zealand agriculture. *New Zealand Journal of Experimental Agriculture*, **6**, 211–15.

Sugiyama, S., Yoneyama, M., Takahashi, N. & Gotoh, K. (1985). Canopy structure and productivity of *Festuca arundinacea* Schreb. swards during vegetative and reproductive growth. *Grass and Forage Science*, **40**, 49–55.

Summerfield, R. J. & Wein, H. C. (1980). Effects of photoperiod and air temperature on growth and yield of economic legumes. In *Advances in Legume Science*, ed. R. J. Summerfield & A. H. Bunting, pp. 17–36. Kew: Royal Botanic Gardens.

Swain, F. G. (1976). Factors affecting pasture establishment in non-cultivated seedbeds on hill country in temperate and subtropical Australia. In *Hill Lands*, ed. J. Luchok, J. D.

Cawthon & M. J. Breslin, pp. 90–100. Morgantown, West Virginia, USA: West Virginia University Press.

Swift, G. & Edwards, R. A. (1983). The effects of spring closing date on the yield and quality of perennial ryegrass and cocksfoot cut for conservation. *Grass and Forage Science*, **38**, 251–60.

Tainton, N. M. (1969). Environmental control of flowering in tropical–subtropical grasses. *Proceedings of the Grassland Society of Southern Africa*, **4**, 49–55.

Tallowin, J. R. B. (1985). Herbage losses from tiller pulling in a continuously grazed perennial ryegrass sward. *Grass and Forage Science*, **40**, 13–18.

Tansley, A. G. (1920). The classification of vegetation and the concepts of development. *Journal of Ecology*, **8**, 118–49.

Taylor, G. B. (1979). The inhibitory effect of light on seed and burr development in several species of *Trifolium*. *Australian Journal of Agricultural Research*, **30**, 895–907.

Taylor, G. B. & Palmer, M. J. (1979). The effect of some environmental conditions on seed development and hardseededness in subterranean clover (*Trifolium subterraneum* L.). *Australian Journal of Agricultural Research*, **30**, 65–76.

Taylor, H. M. & Gardner, H. R. (1963). Penetration of cotton seedling taproots as influenced by bulk density, moisture content, and strength of soil. *Soil Science*, **96**, 153–6.

Taylor, H. M., Jordan, W. R. & Sinclair, T. R. (1983). *Limitations to Efficient Water Use in Crop Production*. 538 pp. Madison: American Society of Agronomy.

Taylor, J. A. (1986). The animal factor in pasture studies. *Proceedings of the XVth International Grassland Congress*, pp. 1140–2. Nishi-nasuno, Japan: The Japanese Society of Grassland Science.

Taylorson, R. B. & Hendricks, S. B. (1977). Dormancy in seeds. *Annual Review of Plant Physiology*, **28**, 331–54.

Teitzel, J. K., Monypenny, J. R. & Rogers, S. J . (1986). Tropical pastures in the farming system: case studies of modelling integration through linear programming. In *Proceedings of the 3rd Australian Conference on Tropical Pastures*, ed. G. R. Murtagh & R. M. Jones, pp. 101–9. Brisbane: Tropical Grassland Society of Australia.

Temple, R. S. & Reh, I. (1984). Livestock populations and factors affecting them. In *Development of Animal Production Systems*, ed. B. Nestel, pp. 33–62. Amsterdam: Elsevier.

Terry, R. A. & Tilley, J. M. A. (1964). The digestibility of the leaves and stems of perennial ryegrass, cocksfoot, timothy, tall fescue, lucerne and sainfoin, as measured by an *in vitro* procedure. *Journal of the British Grassland Society*, **19**, 363–72.

Thomas, C. & Young, J. W. O. (eds.) (1982). *Milk from Grass*. 104 pp. Hurley, UK: Commonwealth Agricultural Bureaux.

Thomas, H. & Stoddart, J. L. (1984). Kinetics of leaf growth in *Lolium temulentum* at optimal and chilling temperatures. *Annals of Botany*, **53**, 341–7.

Thomas, R. G. (1979). The effect of temperature on flowering. *Proceedings of the Agronomy Society of New Zealand*, **9**, 59–64.

Thomas, R. G. (1980). Growth of the white clover plant in relation to seed production. In *Herbage Seed Production*, ed. J. A. Lancashire, pp. 56–63. Palmerston North: New Zealand Grassland Association.

Thompson, K. & Grime, J. P. (1979). Seasonal variation in the seed banks of herbaceous species in ten contrasting habitats. *Journal of Ecology*, **67**, 893–921.

Thompson, P. A. (1981). Ecological aspects of seed germination. *Advances in Research and Technology of Seeds*, **6**, 9–42.

Thornley, J. M. (1976). *Mathematical Models in Plant Physiology*. 318 pp. London: Academic Press.

Thornton, R. F. & Minson, D. J. (1973). The relationship between apparent retention time in the rumen, voluntary intake and apparent digestibility of legume and grass diets in sheep. *Australian Journal of Agricultural Research*, **24**, 889–98.

Thunberg (1793). *Travels in Europe, Africa and Asia performed between the years 1770–79*. Vol. 1. London: Richardson. (Cited by West, 1965.)

Toledo, J. M. (1986a). The role of forage research networking in tropical humid and sub-humid environments. In *Forages in South-East Asian and South Pacific Agriculture*, ed. G. Blair, D. A. Ivory & T. A. Evans, pp. 69–75. Canberra: Australian Centre for International Agricultural Research.

Toledo, J. M. (1986b). Pasture development for cattle production in the major ecosystems of the tropical American lowlands. *Proceedings of the XVth International Grassland Congress*, pp. 74–81. Nishi-nasuno, Japan: The Japanese Society of Grassland Science.

Torssell, B. W. R. (1973). Patterns and processes in the Townsville stylo annual grass pasture ecosystem. *Journal of Applied Ecology*, **10**, 463–78.

Torssell, B. W. R. & McKeon, G. M. (1976). Germination effects on pasture composition in a dry monsoonal climate. *Journal of Applied Ecology*, **13**, 593–603.

Torssell, B. W. R., Rose, C. W. & Cunningham, R. B. (1975). Population dynamics of an annual pasture in a dry monsoonal climate. *Proceedings of the Ecological Society of Australia*, **9**, 157–71.

Tran, V. N. & Cavanagh, A. K. (1984). Structural aspects of dormancy. In *Seed Physiology*, Vol. 2, ed. D. R. Murray, pp. 1–44. Sydney: Academic Press Australia.

Tranter, H. E. & Tranter, R. B. (1981). The demand for amenity grassland. In *Grassland in the British Economy*, ed. J. L. Jollans, pp. 179–207. Reading, UK: Centre for Agricultural Strategy.

Trenbath, B. R. (1978). Models and the interpretation of mixture experiments. In: *Plant Relations in Pastures*, ed. J. R. Wilson, pp. 145–62. Melbourne: Commonwealth Scientific and Industrial Research Organization.

Troughton, A. (1957). The underground organs of herbage grasses. *Commonwealth Agricultural Bureaux Bulletin*, **44**, 163.

Troughton, A. (1977). The rate of growth and partitioning of assimilates in young grass plants: a mathematical model. *Annals of Botany*, **41**, 553–65.

Turkington, R. & Harper, J. L. (1979). The growth, distribution and neighbour relationships of *Trifolium repens* in a permanent pasture. I. Ordination, pattern and contact. *Journal of Ecology*, **67**, 201–18.

Turner, D. W. (1979). Growth and mineral nutrition of the banana – an integrated approach. Unpublished PhD thesis, Macquarie University, Sydney. 259 pp.

Tyler, G. (1971). Distribution and turnover of organic matter and minerals in a shore meadow ecosystem. *Oikos*, **22**, 265–91.

Vallis, I. (1983). Uptake by grass and transfer to soil of nitrogen from ^{15}N labelled legume materials applied to a Rhodes grass pasture. *Australian Journal of Agricultural Research*, **34**, 367–76.

Vallis, I., Harper, L. A., Cathchpoole, V. R. & Weier, K. L. (1982). Volatilization of ammonia from urine patches in a subtropical pasture. *Australian Journal of Agricultural Research*, **33**, 97–107.

Vasil, I. K. (1986). Biotechnology in the improvement of forage crops. In *Proceedings of the XVth International Grassland Congress*, pp. 45–8. Nishi-nasuno, Japan: The Japanese Society of Grassland Science.

Vegis, A. (1964). Dormancy in higher plants. *Annual Review of Plant Physiology*, **15**, 185–224.

Vickery, P. J. (1972). Comparative net primary productivity of grazing systems with different stocking densities of sheep. *Journal of Applied Ecology*, **9**, 307–14.

Vickery, P. J. & Hedges, D. A. (1972). Mathematical relationships and computer routines for a productivity model of improved pasture grazed by merino sheep. *Commonwealth Scientific and Industrial Research Organization Animal Research Laboratory Technical Paper 4*. Canberra: CSIRO.

Vince-Prue, D. (1975). *Photoperiodism in Plants*. 444 pp. London: McGraw-Hill.

Vogel, E. F. de (1980). *Seedlings of Dicotyledons*. 465 pp. Wageningen: Pudoc.

Vorster, M. (1982). The development of the ecological index method for assessing Veld condition in the karoo. *Proceedings of the Grassland Society of Southern Africa*, **17**, 84–9.

Wallace, M. M. H. (1970) Insects of grasslands. In *Australian Grasslands*, ed. R. M. Moore, pp. 361–70. Canberra: Australian National University Press.

Wardlaw, I. F. (1979). The physiological effects of temperature on plant growth. *Proceedings of the Agronomy Society of New Zealand*, **9**, 39–48.

Wareing, P. F. (1982). Hormonal regulation of seed dormancy – past, present and future. In *The Physiology and Biochemistry of Seed Development, Dormancy and Germination*, ed. A. A. Khan, pp. 185–202. Amsterdam: Elsevier Biomedical Press.

Warren-Wilson, J. (1971). Maximum yield potential. In *Transition from Extensive to Intensive Agriculture with Fertilizers. Proceedings of an International Potash Institute Symposium*, pp. 34–56. Berne: International Potash Institute.

Waterman, D. A. (1986). *A Guide to Expert Systems*. Reading, Massachusetts: Addison-Wesley.

Watkin, B. R. & Clements, R. J. (1978). The effects of grazing animals on pastures. In *Plant Relations in Pastures*, ed. J. R. Wilson, pp. 273–90. Melbourne: Commonwealth Scientific and Industrial Research Organization.

Watson, S. J. & Nash, M. J. (1960). *The Conservation of Grass and Forage Crops*. Edinburgh: Oliver and Boyd.

Watt, T. A. & Haggar, R. J. (1980). The effect of defoliation upon yield, flowering and vegetative spread of *Holcus lanatus* growing with and without *Lolium perenne*. *Grass and Forage Science*, **35**, 227–34.

Weaver, J. E. (1926). *Root Development of Field Crops*. 241 pp. New York: McGraw-Hill.

Weeda, W. C. (1967). The effect of cattle dung patches on pasture growth, botanical composition, and pasture utilisation. *New Zealand Journal of Agricultural Research*, **10**, 150–60.

West, O. (1965). *Fire in Vegetation and its use in Pasture Management with Special Reference to Tropical and Subtropical Africa*. 51 pp. CAB Publication 1/1965. Hurley, England: Commonwealth Agricultural Bureaux.

Weston, R. H. (1982). Animal factors affecting feed intake. In *Nutritional Limits to Animal Production from Pastures*, ed. J. B. Hacker, pp. 183–98. Farnham Royal: Commonwealth Agricultural Bureaux.

Wheeler, J. L. (1981a). Field evaluation of complementary forage–pasture systems. In *Forage Evaluation: Concepts and Techniques*, ed. J. L. Wheeler & R. D. Mochrie, pp. 507–16.

Melbourne: Commonwealth Scientific and Industrial Research Organization.

Wheeler, J. L. (1981*b*). Complementing grassland with forage crops. In *Grazing Animals*, ed. F. H. W. Morley, pp. 239–60. Amsterdam: Elsevier.

Wheeler, J. L., Hedges, D. A. & Till, A. R. (1975). A possible effect of cyanogenic glycoside in sorghum on animal requirements for sulphur. *Journal of Agricultural Science*, **84**, 377–9.

Wheeler, J. L., Pearson, C. J. & Robards, G. E. (eds.) (1987). *Temperate Pastures: their Production, Use and Management*. Melbourne: Australian Wool Corporation and Commonwealth Scientific and Industrial Research Organization.

White, D. H., Elliot, B. R., Sharkey, M. J. & Reeves, T. G. (1978). Efficiency of land-use systems involving crop and pastures. *Journal of the Australian Institute of Agricultural Science*, **44**, 21–7.

White, D. J., Wilkinson, J. M. & Wilkins, R. J. (1983). Support energy use in animal production from grassland. In *Efficient Grassland Farming*, ed. A. J. Corrall, pp. 33–42. Hurley, UK: British Grassland Society.

White, J. (1982). The allometric interpretation of the self-thinning rule. *Journal of Theoretical Biology*, **89**, 475–600.

Whitehead, D. C. (1966). *Nutrient Minerals in Grassland Herbage*. Grassland Research Institute, Technical Report 4. 83 pp. Hurley, UK: Commonwealth Agricultural Bureaux.

Whitfield, D. M., Wright, G. C., Gyles, O. A. & Taylor, A. J. (1986). Effects of stage of growth, irrigation frequency and gypsum treatment on CO_2 exchange of lucerne (*Medicago sativa*) grown on a heavy clay soil. *Irrigation Science* (in press).

Whitlam, G. B., Harrison, S. R. & Wilson, J. D. (1970). *Methods of Evaluation of Farm Development Projects*. Technical Bulletin 7. 63 pp. Queensland Department of Primary Industries.

Whittington, W. J. & O'Brien, T. A. (1968). A comparison of yield from plots sown with a single species or a mixture of grass species. *Journal of Applied Ecology*, **5**, 209–13.

Wicklow, D. T. & Zak, J. C. (1983). Viable grass seeds in herbivore dung from a semi-arid grassland. *Grass and Forage Science*, **38**, 25–6.

Wilkins, R. J. (1982). Improving forage quality by processing. In *Nutritional Limits to Animal Production from Pastures*, ed. J. B. Hacker, pp. 389–408. Slough, UK: Commonwealth Agricultural Bureaux.

Wilkinson, J. M. (1983*a*). Silages made from tropical and temperate crops. 1. The ensilage process and its influence on feed value. *World Animal Review*, **45**, 36–42.

Wilkinson, J. M. (1983*b*). Silages made from tropical and temperate crops. Techniques for improving the nutritive value of silage. *World Animal Review*, **46**, 35–40.

Wilkinson, S. R. & Lowrey, R. W. (1973). Cycling of mineral nutrients in pasture ecosystems. In *Chemistry and Biochemistry of Herbage*, Vol. 2, ed. G. W. Butler & R. W. Bailey, pp. 248–316. London: Academic Press.

Williams, C. N. & Biddiscombe, E. F. (1965). Extension growth of grass tillers in the field. *Australian Journal of Agricultural Research*, **16**, 14–22.

Williams, E. D. (1984). Some effects of fertilizer and frequency of defoliation on the botanical composition and yield of permanent grassland. *Grass and Forage Science*, **39**, 311–15.

Williams, O. B. (1964). Energy flow and nutrient cycling in ecosystems. *Proceedings of the Australian Society of Animal Production*, **5**, 291–300.

Williams, O. B. (1970). Population dynamics of two perennial grasses in Australian semi-arid grassland. *Journal of Ecology*, **58**, 869–75.

Wilman, D. & Asiegbu, J. E. (1982). The effect of clover variety, cutting interval and nitrogen application on herbage yields, proportions and heights in perennial ryegrass – white clover swards. *Grass and Forage Science*, **37**, 1–13.

Wilman, D. & Mohamed, A. A. (1980). Early spring and late autumn response to applied nitrogen in four grasses. 2. Leaf development. *Journal of Agricultural Science*, **94**, 443–53.

Wilson, B. (1984). *Systems: Concepts, Methodologies and Applications*. 339 pp. Chichester, UK: John Wiley & Sons.

Wilson, J. R. (ed.) (1978). *Plant Relations in Pastures*. Melbourne: Commonwealth Scientific and Industrial Research Organization.

Wilson, J. R. (1982). Environmental and nutritional factors affecting herbage quality. In *Nutritional Limits to Animal Production from Pastures*, ed. J. B. Hacker, pp. 111–31. Farnham Royal: Commonwealth Agricultural Bureaux.

Wilson, J. R. (1983). Effects of water stress on *in vitro* dry matter digestion and chemical composition of herbage of tropical pasture species. *Australian Journal of Agricultural Research*, **34**, 377–90.

Wilson, J. R. (1986). An interdisciplinary approach for increasing yield and improving quality of forages. *Proceedings of the XVth International Grassland Congress*, pp. 49–55. Nishi-nasuno, Japan: The Japanese Society of Grassland Science.

Wilson, J. R., Brown, R. H. & Windham, W. R. (1983). Influence of leaf anatomy on the dry matter digestibility of C4, C3 and C3/C4 intermediate types of *Panicum* species. *Crop Science*, **23**, 141–6.

Wilson, J. R. & Hattersley, P. W. (1983). *In vitro* digestion of bundle sheath cells in rumen fluid and its relation to the suberized lamella and C4 photosynthetic type in *Panicum* species. *Grass and Forage Science*, **38**, 219–23.

Wilson, J. R. & Ludlow, M. M. (1983). Time trends for change in osmotic adjustment and water relations of leaves of *Cenchrus ciliaris* during and after water stress. *Australian Journal of Plant Physiology*, **10**, 15–24.

Wilson, J. R., Ludlow, M. M., Fisher, M. J. & Schultz, E. D. (1980). Adaption to water stress of the leaf water relations of four tropical forage species. *Australian Journal of Plant Physiology*, **7**, 207–20.

Wilson, J. R. & Minson, D. J. (1980). Prospects for improving the digestibility and intake of tropical grasses. *Tropical Grasslands*, **14**, 253–9.

Wilson, J. R. & Wong, C. C. (1982). Influence of shading on the nutritive quality of green panic and Siratro. *Australian Journal of Agricultural Research*, **33**, 937–49.

Wimaladharma, K. P. (1985). The impact of sociology on livestock production management. *Proceedings of Regional Workshop on Livestock Production Management*, pp. 95–126. Manila: Asian Development Bank.

Winks, L. W. (1984). Cattle growth in the dry tropics of Australia. *Australian Meat Research Committee Review*, **45**, 1–44.

Wit, C. T., de, Tow, P. G. & Ennik, G. C. (1966). Competition between legumes and grasses. *Versl. Landbouwk. Onderz.* **687**, 3–30.

Witschi, P. A. & Michalk, D. L. (1979). The effect of sheep treading and grazing on pasture and soil characteristics of irrigated annual pastures. *Australian Journal of Agricultural Research*, **30**, 741–50.

Woledge, J. (1978). The effect of shading during vegetative and reproductive growth on the photosynthetic capacity of leaves in a grass sward. *Annals of Botany*, **42**, 1085–9.

Woledge, J. (1979). Effect of flowering on the photosynthetic capacity of ryegrass leaves grown with and without natural shading. *Annals of Botany*, **44**, 197–207.

Wolfe, E. C. (1981). *Adaptation of Subterranean Clover to Moisture Stress*. 48 pp. Sydney: Department of Agriculture, NSW.

Wolfe, E. C. & Lazenby, A. (1973). Grass-white clover relationships during pasture development 2. *Australian Journal of Experimental Agriculture and Animal Husbandry*, **13**, 575–80.

Wolton, K. M., Brockman, J. S. & Shaw, P. G. (1970). The effect of stage of growth at defoliation on white clover in mixed swards. *Journal of the British Grassland Society*, **25**, 113–18.

Woodman, H. E. & Norman, D. B. (1932). Nutritive value of pasture. IX. The influence of the intensity of grazing on nutritive value of pasture herbage. *Journal of Agricultural Science*, **22**, 852–73.

Woodward, R. G. & Morley, F. H. W. (1974). Variation in Australian and European collections of *Trifolium glomeratum* L. and the provisional distribution of the species in southern Australia. *Australian Journal of Agricultural Research*, **25**, 73–88.

Yates, J. J. (1982). A technique for estimating rate of disappearance of dead herbage from pasture. *Grass and Forage Science*, **37**, 249–52.

Yoda, K., Kira, T., Ogawa, H. & Hozumi, K. (1963). Self-thinning in overcrowded pure stands under cultivated and natural conditions. *Journal of Biology of Osaka City University*, **14**, 107–29.

INDEX

Italicized page numbers indicate those pages on which definitions appear.